Y0-AAJ-247

Hadron Physics
at Very High Energies

FRONTIERS IN PHYSICS

David Pines, Editor

S. L. Adler and R. F. Dashen
Current Algebras and Applications to Particle Physics, 1968

P. W. Anderson *Concepts in Solids: Lectures on the Theory of Solids, 1965 (2nd printing, 1971)*

V. Barger and D. Cline
Phenomenological Theories of High Energy Scattering: An Experimental Evaluation, 1967

N. Bloembergen *Nonlinear Optics: A Lecture Note and Reprint Volume, 1965*

N. Bloembergen *Nuclear Magnetic Relaxation: A Reprint Volume, 1961*

R. Brout *Phase Transitions, 1965*

G. F. Chew *S-Matrix Theory of Strong Interactions: A Lecture Note and Reprint Volume, 1961*

P. Choquard *The Anharmonic Crystal, 1967*

P. G. de Gennes *Superconductivity of Metals and Alloys, 1966*

R. P. Feynman *Quantum Electrodynamics: A Lecture Note and Reprint Volume, 1961 (2nd printing, with corrections, 1962)*

R. P. Feynman *The Theory of Fundamental Processes: A Lecture Note Volume, 1961*

R. P. Feynman *Statistical Mechanics: A Set of Lectures, 1972*

R. P. Feynman *Photon-Hadron Interactions, 1972*

H. Frauenfelder *The Mössbauer Effect: A Review—with a Collection of Reprints, 1966*

S. C. Frautschi *Regge Poles and S-Matrix Theory, 1963*

H. L. Frisch and J. L. Lebowitz
The Equilibrium Theory of Classical Fluids: A Lecture Note and Reprint Volume, 1964

M. Gell-Mann and Y. Ne'eman
The Eightfold Way: (A Review—with a Collection of Reprints), 1964

W. A. Harrison *Pseudopotentials in the Theory of Metals, 1966 (2nd printing, 1971)*

R. Hofstadter — *Electron Scattering and Nuclear and Nucleon Structure: A Collection of Reprints with an Introduction, 1963*

D. Horn and F. Zachariasen
Hadron Physics at Very High Energies, 1973

M. Jacob and G. F. Chew
Strong-Interaction Physics: A Lecture Note Volume, 1964

L. P. Kadanoff and G. Baym
Quantum Statistical Mechanics: Green's Function Methods in Equilibrium and Nonequilibrium Problems, 1962 (2nd printing, 1971)

I. M. Khalatnikov
An Introduction to the Theory of Superfluidity, 1965

J. J. J. Kokkedee
The Quark Model, 1969

A. M. Lane — *Nuclear Theory: Pairing Force Correlations to Collective Motion, 1963*

T. Loucks — *Augmented Plane Wave Method: A Guide to Performing Electronic Structure Calculations—A Lecture Note and Reprint Volume, 1967*

A. B. Migdal and V. Krainov
Approximation Methods in Quantum Mechanics, 1969

A. B. Migdal — *Nuclear Theory: The Quasiparticle Method, 1968*

Y. Ne'eman — *Algebraic Theory of Particle Physics: Hadron Dynamics in Terms of Unitary Spin Currents, 1967*

P. Nozières — *Theory of Interacting Fermi Systems, 1964*

R. Omnès and M. Froissart
Mandelstam Theory and Regge Poles: An Introduction for Experimentalists, 1963

D. Pines — *The Many-Body Problem: A Lecture Note and Reprint Volume, 1961*

R. Z. Sagdeev and A. A. Galeev
Nonlinear Plasma Theory, 1969

J. R. Schrieffer — *Theory of Superconductivity, 1964 (2nd printing, 1971)*

J. Schwinger — *Quantum Kinematics and Dynamics, 1970*

FRONTIERS IN PHYSICS

David Pines, Editor

E. J. Squires — *Complex Angular Momenta and Particle Physics: A Lecture Note and Reprint Volume, 1963*

L. Van Hove, N. M. Hugenholtz, and L. P. Howland — *Problems in Quantum Theory of Many-Particle Systems: A Lecture Note and Reprint Volume, 1961*

IN PREPARATION:

E. R. Caianiello — *Combinatorics and Renormalization in Quantum Field Theory*

S. Doniach and E. H. Sondheimer — *Green's Functions for Solid State Physicists*

S. Ichimaru — *Basic Principles of Plasma Physics: A Statistical Approach*

G. E. Pake and T. L. Estle — *The Physical Principles of Electron Paramagnetic Resonance, 2nd edition, completely revised, enlarged, and reset*

Hadron Physics at Very High Energies

David Horn
Tel-Aviv University

Fredrik Zachariasen
California Institute of Technology

1973
W. A. BENJAMIN, INC.
ADVANCED BOOK PROGRAM
Reading, Massachusetts

London · Amsterdam · Don Mills, Ontario · Sydney · Tokyo

Library of Congress Cataloging in Publication Data

Horn, David, 1937-
Hadron physics at very high energies.

(Frontiers in physics)
1. Hadrons. 2. Nuclear reactions.
I. Zachariasen, Fredrik, joint author. II. Title.
III. Series.
QC793.5.H328H67 539.7'216 73-9616

Reproduced by W. A. Benjamin, Inc., Advanced Book Program, Reading, Massachusetts, from camera-ready copy prepared by the authors.

Manufactured in the United States of America

ISBN 0-8053-4402-0 (hardbound)
ISBN 0-8053-4403-9 (paperback)
ABCDEFGHIJ-MA-79876543

Contents

EDITOR'S FOREWORD

The problem of communicating in a coherent fashion the recent developments in the most exciting and active fields of physics seems particularly pressing today. The enormous growth in the number of physicists has tended to make the familiar channels of communication considerably less effective. It has become increasingly difficult for experts in a given field to keep up with the current literature; the novice can only be confused. What is needed is both a consistent account of a field and the presentation of a definite "point of view" concerning it. Formal monographs cannot meet such a need in a rapidly developing field, and, perhaps more important, the review article seems to have fallen into disfavor. Indeed, it would seem that the people most actively engaged in developing a given field are the people least likely to write at length about it.

FRONTIERS IN PHYSICS has been conceived in an effort to improve the situation in several ways. One of these is to take advantage of the fact that the leading physicists today frequently give a series of lectures, a graduate seminar, or a graduate course in the special fields of interest. Such lectures serve to summarize the present status of a rapidly developing field and may well constitute the only coherent account available at the time. Often, notes on lectures exist (prepared by the lecturer himself, by graduate students, or by postdoctoral fellows) and have been distributed in mimeographed form on a limited basis. One of the principal purposes of the FRONTIERS IN PHYSICS Series is to make such notes available to a wider audience of physicists.

It should be emphasized that lecture notes are necessarily rough and informal, both in style and content, and those in the series will prove no exception. This is as it should be. The point of the series is to offer new, rapid, more informal, and it is hoped, more effective ways for physicists to teach one another. The point is lost if only elegant notes qualify.

A second way to improve communication in very active fieldsof physics is by the publication of collections of reprints of recent articles. Such collections are themselves useful to people working in the field. The value of the reprints would, however, seem much enhanced if the collection would be accompanied by an introduction of moderate length, which would serve to tie the collection together and, necessarily, constitute a brief survey of the present status of the field. Again, it is appropriate that such an introduction be informal, in keeping with the active character of the field.

A third possibility for the series might be called an informal monograph, to connote the fact that it represents an intermediate step between lecture notes and formal monographs. It would offer the author an opportunity to present his views of a field that has developed to the point at which a summation might prove extraordinarily fruitful, but for which a formal monograph might not be feasible or desirable.

Fourth, there are the contemporary classics--papers or lectures which constitute a particularly valuable approach to the teaching and learning of physics today. Here one thinks of fields that lie at the heart of much of present-day research, but whose essentials are by now well understood, such as quantum electrodynamics or magnetic resonance. In such fields some of the best pedagogical material is not readily available, either because it consists of papers long out of print or lectures that have never been published.

The above words, written in August, 1961, seem equally applicable today (which may tell us something about developments in communication in physics during the past 12 years). David Horn and Fredrik Zachariase have contributed significantly to the studies of had physics at very high energies. It now gives me great pleasure to welcome them as contributors to the FRONTIERS IN PHYSICS Series. May their volume play well its part in the continuing effort to keep abreas of developments all investigators and students of particle physics.

DAVID PINES

Urbana, Illinois

Summer 1973

Preface

For years the belief has been widespread that the physics of the strongly interacting particles becomes simple at asymptotic energies. At energies up to several GeV, cross sections display a complicated structure reflecting the existence of numerous resonances; there are dips and wiggles in angular distributions; polarizations change signs and change signs again; and in general the physics is determined by many competing effects which it is very difficult to disentangle theoretically. At very high energies, in contrast, the hope has been that cross sections become smooth and that most phenomena tend to a sort of steady state in which the underlying physics can be plainly seen, and theoretical understanding becomes possible.

The only natural dimensions with which we are familiar in particle physics are of the order of one GeV; hence, it has been anticipated that the asymptotic regime would be attained at energies above this, perhaps at energies of a few hundred GeV. With the advent of the CERN Intersecting Storage Rings and the NAL accelerator, this regime has finally become accessible experimentally.

We are therefore at long last in the situation of having a great deal of experimental information rapidly becoming available at energies where theoretical interpretation of it may be feasible. And it indeed seems to be the case that much simplicity and order does exist in the data. To be sure, we do not yet have a detailed theory in terms of which everything can be understood quantitatively; yet many rules are becoming evident, and parts of the data are being correlated in terms of various theoretical models some of which will, no doubt, be incorporated into the eventual complete theory.

It would seem, then, that this is a useful time to summarize the data and the models we now have for correlating and interpreting it. This is what we intend to do here. The presentation is directed at physicists desiring a survey of this field, and at advanced graduate students. Familiarity with conventional field theory and with two-body Regge pole phenomenology is assumed, but within these limits we have tried to provide a self-contained discussion of very high energy hadronic processes of sufficient depth to enable the reader to follow in some detail both experimental and theoretical progress in this field.

It is a pleasure to thank Mrs. A. Lotspeich for typing the book and Mrs. H. Tuck for typing the draft and assisting in the editing of the final version. One of us (D. H.) would like to thank the California Institute of Technology for its hospitality. Finally, thanks are due to all our colleagues for stimulating discussions.

Pasadena, California
April 1973

David Horn
Fredrik Zachariasen

Introduction

Our intent is to present an up-to-date review of the major experimental facts, phenomenological analyses and theoretical ideas which make up our present understanding of hadronic interactions at very high energies. The order of the presentation is as follows. First, a survey of the experimental situation, with the data summarized in a set of rules which seem to be obeyed by all results on total cross sections, one- and two-body inclusive cross sections, and two-body and many-body exclusive cross sections. Next, a description and interpretation of these same phenomena in the context of our most successful high energy phenomenology to date; namely, Regge poles. Finally, an outline of those theoretical models which we feel contain some ideas or points of view which have a chance of surviving and contributing to a complete theory, when and if a complete theory is found.

There are also a number of basic theoretical results relevant to high energy phenomena. These we have located in a series of appendices to avoid breaking up the continuity of the text itself.

We shall use the following notation, normalization, and definitions throughout.

A high energy exclusive reaction in which a collision between particles A and B produces particles C_1 .. C_n is written $A + B \to C_1 + C_2 + .. + C_n$; an inclusive reaction in which any other particles are produced in addition to C_1 .. C_n is written either as $A + B \to C_1 + .. + C_n + X$, or as $(AB, C_1 \ldots C_n)$. The 4-momenta of A and B are called p_1 and p_2; those of C_1 .. C_n are called q_1 .. q_n. The energies of A and B are E_1 and E_2, and of C_1 .. C_n are ω_1 .. ω_n. The components of any $\vec{q}$ parallel or transverse to the collision axis are denoted q_L and q_T respectively. We shall usually suppress spin indices. We also make use of the usual relativistic scalar $s = (p_1 + p_2)^2 \equiv P^2$, the square of the total c.m. energy, and we shall always be concerned with situations in which s is much larger than the masses $m_A{}^2$, $m_B{}^2$, $m_{C_1}{}^2$ .. $m_{C_n}{}^2$ of all particles.

Our transition amplitude $T_{2\to n}\,(p_1 p_2 \to q_1 .. q_2)$ describing the reaction $A + B \to C_1 + .. + C_n$ is normalized through the definition

$$S_{2\to n}(p_1 p_2 \to q_1 .. q_n) = \delta_{n2}\, \delta_{p_1 q_1}\, \delta_{p_2 q_2} + i\,(2\pi)^4 \delta^{(4)}(P - q_1 - .. - q_n) \frac{T_{2\to n}(p_1 p_2 \to q_1 .. q_n)}{\sqrt{2E_1 2E_2 2\omega_1 2\omega_2 .. 2\omega_n}} \qquad \text{(I.1)}$$

where S is the S-matrix. From this it follows that the total cross section for A+B→X is

$$\sigma_{AB} = \frac{1}{2s}\sum_{n=2}^{\infty} \int \frac{d^3 q_1}{(2\pi)^3} \cdot\cdot \int \frac{d^3 q_n}{(2\pi)^3} \frac{1}{2\omega_1} \cdot\cdot \frac{1}{2\omega_n} \cdot$$

$$\cdot\ (2\pi)^4 \delta^{(4)} (P - q_1 - \cdots - q_n)$$

$$\cdot\ |T_{2\to n}\ (p_1 p_2 \to q_1 \cdot\cdot q_n)|^2 \qquad \text{(I.2)}$$

This we shall often abbreviate to

$$\sigma = \frac{1}{2s}\sum_{n=2}^{\infty} \int d\Phi_n\ |T_{2\to n}|^2 \qquad \text{(I.3)}$$

with the definition of the element of phase space

$$d\Phi_n = \frac{d^3 q_1}{(2\pi)^3} \cdot\cdot \frac{d^3 q_n}{(2\pi)^3} \frac{1}{2\omega_1} \cdot\cdot \frac{1}{2\omega_n} (2\pi)^4 \cdot$$

$$\cdot\ \delta^{(4)} (P - \Sigma q) \qquad \text{(I.4)}$$

Similarly, the one particle inclusive cross section for (AB, C) is

$$\frac{d\sigma_{AB}^{C}}{d^3 q} = \frac{1}{2s} \sum_{n=2}^{\infty} \int d\Phi_n \sum_{i_C=1}^{n_C} \delta^{(3)}(\vec{q} - \vec{q}_{i_C}) |T_{2\to n}|^2 \qquad \text{(I.5)}$$

where i_c sums over the particles of type C in n. Frequently we will use the relativistically invariant cross sections

$$\rho_{AB}^{C} \equiv \omega_C \frac{d\sigma_{AB}^{C}}{d^3 q}$$

to describe one particle inclusive processes.

The two particle inclusive cross section for $A+B\to C_1+C_2+X$ is

$$\frac{d\sigma_{AB}^{C_1C_2}}{d^3q_1\, d^3q_2} = \frac{1}{2s}\sum_n \int d\Phi_n \sum_{i_{C_1}=1}^{n_{C_1}} \delta^{(3)}\left(\vec{q}_1 - \vec{q}_{i_{C_1}}\right) \cdot$$

$$\cdot \sum_{i_{C_2}=1}^{n_{C_2}} \delta^{(3)}\left(\vec{q}_2 - \vec{q}_{i_{C_2}}\right) |T_{2\to n}|^2 \qquad \text{(I.6)}$$

and so on.

The exlusive two particle cross section for $A + B \to C_1 + C_2$ is

$$\frac{d\sigma_{AB\to C_1+C_2}}{dt} = \frac{1}{16\pi s^2} |T_{AB\to C_1C_2}(p_1p_2\to q_1q_2)|^2. \qquad \text{(I.7)}$$

Finally, from (I.1) and the unitarity of the S matrix we also have the unitarity relation for T: The extension of (I.1) to any T matrix element T_{fi} connecting an arbitrary state i to an arbitrary state f is

$$S_{fi} = \delta_{fi} + i\,(2\pi)^4\delta^{(4)}(P_f - P_i)\frac{T_{fi}}{\sqrt{\prod_i 2\omega \prod_f 2\omega}} \qquad \text{(I.8)}$$

where P_f and P_i are the total 4-momenta of the states, and $\prod_i 2\omega$ means the product of twice the energy over all particles in state i.

Then $$\sum_n S_{fn}\, S^+_{ni} = \delta_{fi} \qquad \text{(I.9)}$$

implies that

$$T_{fi} - T_{if}^{*} = i \sum_n T_{fn} \ (2\pi)^4 \frac{\delta^{(4)}(p_n - p_i)}{(\Pi 2\omega)_n} T_{in}^{*} \qquad (I.10)$$

The sum over a set of states n is

$$\sum_n = \sum_n \int \frac{d^3 q_1}{(2\pi)^3} \cdots \int \frac{d^3 q_n}{(2\pi)^3}$$

with the sum over spin indices implied.

For the elastic amplitude, eq. (I.10) reduces to

$$A(s,\ t) = \frac{1}{2} \sum_n \int d\Phi_n \ T_{2\to n}(p_1 p_2 \to q_1 \cdot\cdot q_n) \cdot$$

$$\cdot T_{2\to n}^{*} \ (p_1' \ p_2' \to q_1 \cdot\cdot q_n) \qquad (I.11)$$

where p_1 and p_2 are the two initial momenta while p_1' and p_2' are the two final ones. Thus $s = (p_1 + p_2)^2$ and $t = (p_1' - p_1)^2$, and, of course, $p_1 + p_2 = p_1' + p_2'$.

We define s-channel partial wave amplitudes for elastic scattering by

$$T(s,j) = \frac{1}{2} \int_{-1}^{1} dx \ P_j(x_s) \ T(s,t) \qquad (I.12)$$

for the case of spinless particles, with $x_s = \cos\theta_s = 1 + 2t/s$. Thus we have

$$T(s,t) = \sum_j \ (2j+1) \ P_j(x_s) \ T(t,j) \qquad (I.13)$$

and from (I.11) it follows that we can represent $T(t,j)$ as

$$T(t,j) = 16\pi \, \frac{\eta_j(s)e^{2i\delta_j(s)} - 1}{2i} \tag{I.14}$$

in terms of a phase shift $\delta_j(s)$ and an absorption $\eta_j(s)$. We have $0 \leq \eta \leq 1$.

The impact parameter representation of the elastic amplitude is defined through the Bessel transform

$$T(s,b) = \int_0^\infty \sqrt{-t}\; d\sqrt{-t} \;\; J_0(b\sqrt{-t})\; T(s,t) \tag{I.15}$$

so that

$$T(s,t) = \int_0^\infty bdb \; J_0(b\sqrt{-t})\; T(s,b). \tag{I.16}$$

Comparing this with (I.14) tells us to write

$$T(s,b) = 8\pi s \left(\frac{\eta(s,b)\; e^{2i\delta(s,b)} - 1}{2i} \right) \tag{I.17}$$

where now η and δ are the absorption and phase shift as functions of impact parameter. We may also define the eikonal

$$\chi(s,b) = \delta(s,b) - i/2 \; \ell n \; \eta(s,b) \tag{I.18}$$

and hence

$$T(s,t) = 8\pi s \int_0^\infty bdb \; J_0(b\sqrt{-t}) \left(\frac{e^{2i\chi(s,b)} - 1}{2i} \right) \tag{I.19}$$

The translation of the unitarity relation (I.11) into partial wave and impact parameter language is given in appendix D.

A number of useful reviews of many of the topics presented here have recently appeared in the literature. Among them are:

E.L. Berger, Colloquim on Multiparticle Dynamics, Helsinki, 1971.
W.R. Frazer et al, Rev. Mod. Phys. 44 284, (1972)
G. Giacomelli, Rapporteur talk, Chicago Conference (1972), Vol. 3, p. 219.
D. Horn, Physics Reports 4C 1, (1972).
M. Jacob, Rapporteur talk, Chicago Conference (1972), Vol. 3, p. 373.
D.R.O. Morrison, Review talk, Oxford Conference (1972).
L. Van Hove, Physics Reports 1C, 347, (1971).

PART I

PHENOMENOLOGY

1

Total and Topological Cross Sections

Total cross sections provide us with our first clues to the structure of hadrons and to their interactions. They involve the summation over all outgoing channels in the production process A + B → anything, and are defined by

$$\sigma_T(s) = \frac{1}{2s} \sum_{n=2}^{\infty} \int \prod_{i=1}^{n} \frac{d^3 q_i}{(2\pi)^3 2\omega_i} (2\pi)^4 \cdot$$

$$\cdot \delta^{(4)}\left(P - \sum_{j=1}^{n} q_j\right) \left| T(p_1 + p_2 \rightarrow q_1 + \cdots + q_n) \right|^2 \tag{1.1}$$

where an explicit sum over all phase space of T-matrix elements (and an implicit sum over spin states) is performed. In looking at this equation one may be tempted to think that the total cross-sections conceal more than they reveal.

Unitarity supplies us with the optical theorem which states that (see Appendix A)

$$s\,\sigma_T^{AB}(s) = \mathrm{Im}\,T_{AB\to AB}(s, t = o) \quad (1.2)$$

thus relating the total cross section to a two-body scattering amplitude. This direct relation of a measured experimental number to a scattering amplitude is quite unique. It can be contrasted with a differential cross section for AB→CD which is related to the sum of the absolute squares of the transition amplitudes in all helicity states. This unique feature comes from the summation over "anything" and, in modern terminology,[1)] has acquired the name "inclusive process." The sum over all states plays the role of a unit operator in quantum mechanics and enables one to obtain eq.(1.2).

The magnitude of the total cross section reflects the range of the forces that act between the particles. This effective range determines the size of their interaction region, and this is the measured quantity, as is implied by the name "cross section." The fact that the hadronic total cross sections are of the order of tens of mb leads us to think of them as objects with the characteristic dimension of a fermi. The first question we have, then, to ask ourselves is to what extent such a picture is energy independent. This question and its various implications, will accompany us throughout this book. The experimental answers that have become available recently show a very smooth behavior with at most small changes in σ_T as s is changed by almost two orders of magnitude.

Total cross sections in the energy range below $s \approx 60\ \mathrm{GeV}^2$ show either a constant or a decreasing energy behavior (see fig. 1.1). They can then be simply parametrized by $\sigma_T \approx A + Bs^{\alpha-1}$ with $\alpha \approx \frac{1}{2}$.[2] The constants A and B are connected to the quantum numbers of the direct channel. Thus the fact that $\sigma_T(K^+p)$ is constant whereas other meson-baryon cross-sections are decreasing may be correlated with the observation that this channel is exotic (i.e. has quantum numbers different from those of resonances in a quark model). This observation leads to a two-component interpretation of the scattering cross sections.[3] One component is connected to direct channel resonances, and its contribution to the cross-section is decreasing with energy. Direct channel resonances can be clearly distinguished only below $s \approx 4\ \mathrm{GeV}^2$, and at higher energies this component is described by Regge exchanges in the t-channel, which are said to be dual to the s-channel resonances. This duality is based on an analysis in terms of finite energy sum rules that serve to define its mathematical meaning.[4]

The second component of scattering is identified with the nonresonant background in the s-channel (including, for instance, production of two resonances). This term is the one responsible for diffraction scattering and stays roughly constant over the whole measured energy range. When described in a Regge language it is called the Pomeron (or the Pomeranchuk singularity). The nature of the Pomeron is our subject. The Serpukhov data changed[5)] the simple parameterization of the cross sections mentioned above. They showed that the decreasing trend disappears and turns into a constant or even an increasing one. In particular, we note that the K^+p cross-section has a clear rising trend above $s \approx 50\ \text{GeV}^2$ (or $p \approx 25$ GeV in fig. 1.1). A similar result has recently been found in pp cross sections measured in the CERN Intersecting Storage Rings (ISR). There one studies an energy range up to $s \approx 3000\ \text{GeV}^2$ and one finds[6)] evidence for an increase of $\sigma_T(pp)$ starting at about $s \approx 500\ \text{GeV}^2$, as seen in fig. 1.2. With the advent of the National Accelerator Laboratory (NAL) at Batavia, Ill., we may hope to soon have a detailed picture of all other meson-nucleon cross sections in the range $s \lesssim 800\ \text{GeV}^2$. In particular, the careful study of $\sigma_T(K^+p)$ and $\sigma_T(pp)$ may allow us to determine the functional form of the s-dependence. Even then it is still unclear whether we are already in some asymptotic region or whether we still see transitionary effects. We will therefore discuss models with asymptotic constant or even decreasing cross sections, as well as those that lead to logarithmic increases. Although the latter seem to be favored by recent data, the only general statement that we can safely make is that

Rule 1. σ_T are roughly constant or slowly

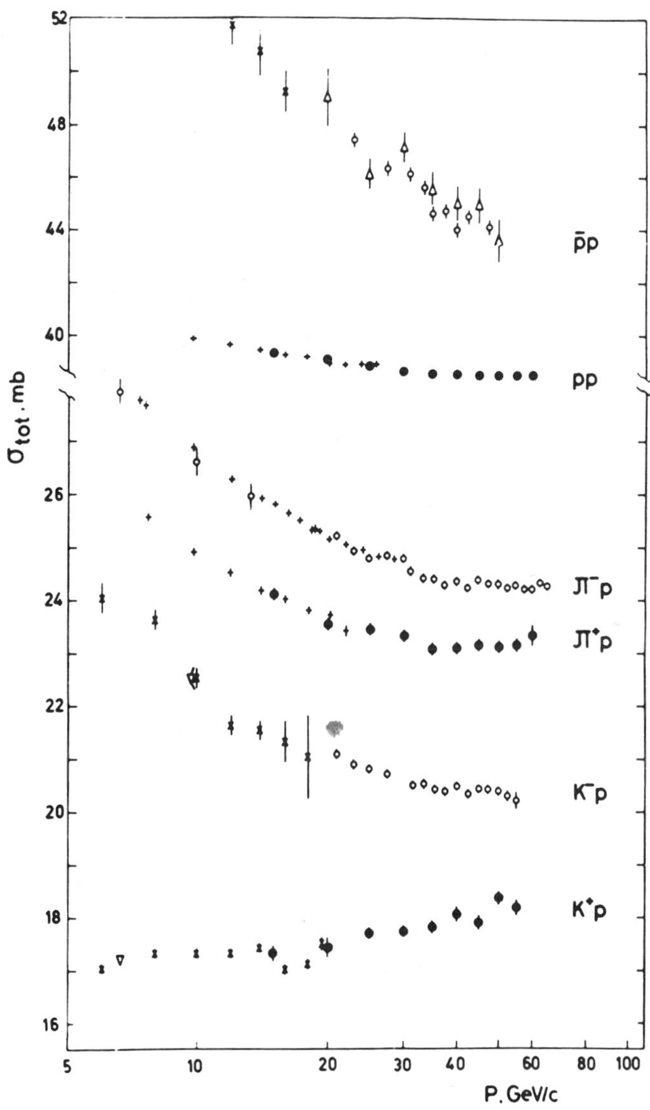

Fig. 1.1. Total cross sections plotted vs. the laboratory momentum $p \approx s/2m_p$. Included are data points from the Serpukhov accelerator. This figure is taken from ref. 5.

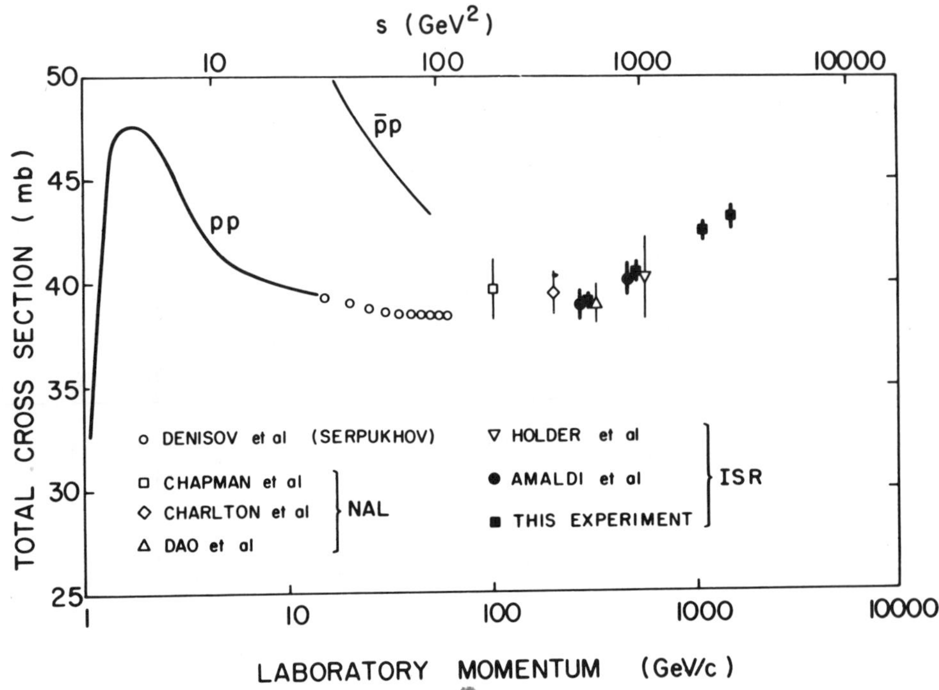

Fig. 1.2. Recent pp total cross section data in the ISR energy range. Figure taken from ref. 6. The new points correspond to an extrapolation of elastic scattering data to the optical points.

changing with s. This is in accord with the Froissart bound (see Appendix B) which states that $\sigma_T \lesssim \ln^2 s$.

There are then several basic properties of the Pomeron that may be seen in fig. 1.1. At finite energies, all are qualitative and approximate since the elimination of the lower Regge terms is model dependent and so is also the energy dependence of the Pomeron itself. Nevertheless, we see that characteristically

Rule 2. $\sigma_T(AB) = \sigma_T(\bar{A}B)$. This is known as the Pomeranchuk theorem (see Appendix C). It can be derived from unitarity and analyticity if the cross section increases; otherwise one has to add an assumption about the real part which is tantamount to assuming the existence of the theorem. Another statement that is often included in the same category is

Rule 3. $\sigma_T(A_iB_j) = \sigma_T(A_mB_n)$, where A_i and A_m as well as B_j and B_n belong to the same isospin multiplets respectively. This is equivalent to saying that the Pomeron is an isospin singlet exchange. One may wonder if the Pomeron is perhaps also an SU_3 singlet. The behavior of the data below $s \approx 60$ GeV^2 does not look like that, but with the new trends observed in the Serpukhov energy range this possibility cannot be excluded.

It is interesting to ask how this constant or nearly constant cross section is built up. σ_T is composed of many partial cross-sections, which, individually, mostly die out rapidly as a function of energy (diffractive production is an exception and will be discussed later). These may nevertheless add up to a nearly constant total cross-section because of the ever increasing number of channels that open as the energy grows. For example, if we look at topological*) cross sections σ_n for the production of n charged particles, which still include the summation over many different channels, clear energy variations can be seen (fig. 1.3). All these variations match each

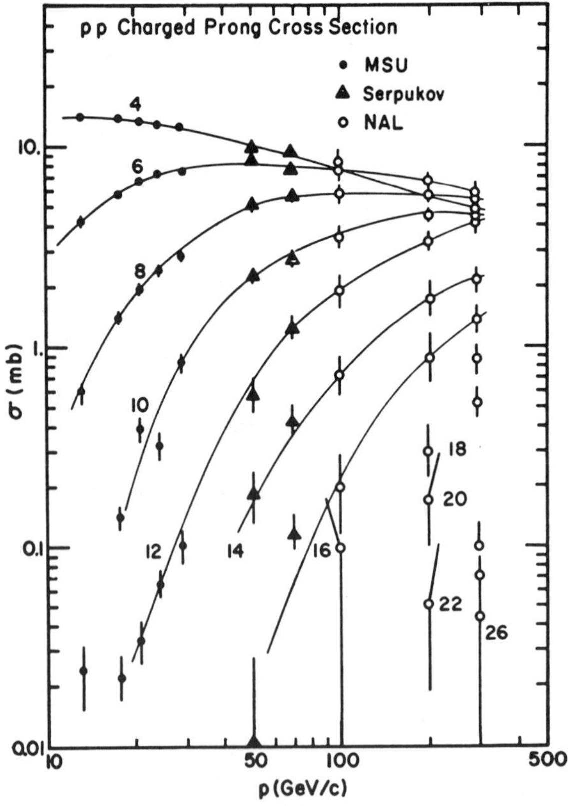

Fig. 1.3. Variation of topological cross sections with incoming lab. momentum. Figure includes the recent NAL results, and is taken from ref.7.

---

* The word topological is used because in bubble chambers these processes show up as having n outgoing prongs.

other so as to build a $\sigma_T = \sum_n \sigma_n$ which is approximately constant. The sum has nonuniform characteristics and, as the energy grows, the σ_n pattern shifts and higher n values become more and more important. As can be seen from fig. 1.3 this variation occurs on a $\ell n\ s$ scale. Therefore one also finds that the average multiplicity

$$\langle n\rangle = \frac{1}{\sigma_T} \sum_n n\ \sigma_n \tag{1.3}$$

grows slowly with s, as shown in fig. 1.4a. It is more common to use the definition

$$\langle n\rangle = \frac{1}{\sigma_I} \sum_n n\ \sigma_n \tag{1.4}$$

instead of (1.3). The difference is that now one excludes from σ_T the elastic contribution and normalizes to the inelastic cross section, given by

$$\sigma_I = \sigma_T - \sigma_{e\ell} \tag{1.5}$$

The data shown in fig. 1.4 are based on the definition (1.4). With either definition one finds that $\langle n\rangle$ grows slowly with s. As we will see, there are theories that lead us to expect a logarithmic dependence, and this may indeed be the case experimentally, although some power behavior cannot be ruled out by the data.

Next let us look at a plot of σ_n vs. n. It is customary to compare this curve with the Poisson distribution that would result from a model of

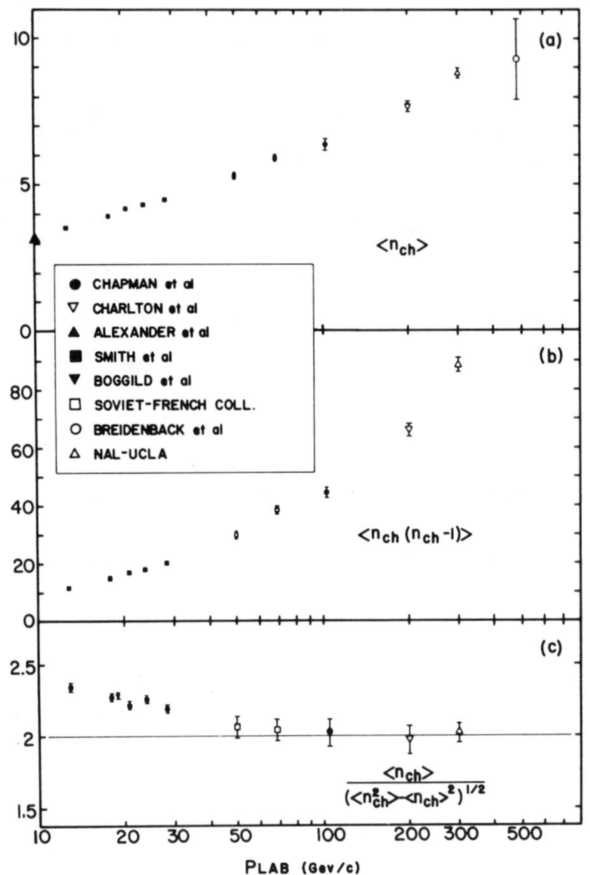

Fig. 1.4. Experimental results for averages over inelastic prong distributions in pp reactions. Recent NAL data are shown at p=100, 200, and 300 GeV/c. Taken from ref. 8.

independent particle production (fig. 1.5). The pp data show a pattern that is wider than a Poisson distribution. In fact, defining the width (or dispersion) of the distribution by

$$D^2 = \langle n^2\rangle - \langle n\rangle^2 \tag{1.6}$$

one finds that all the recent pp data are consistent with a phenomenological fit[10)] (see fig. 1.4c).

$$\langle n\rangle \simeq 2D \tag{1.7}$$

to be contrasted with the $\langle n\rangle = D^2$ property of a Poisson distribution.

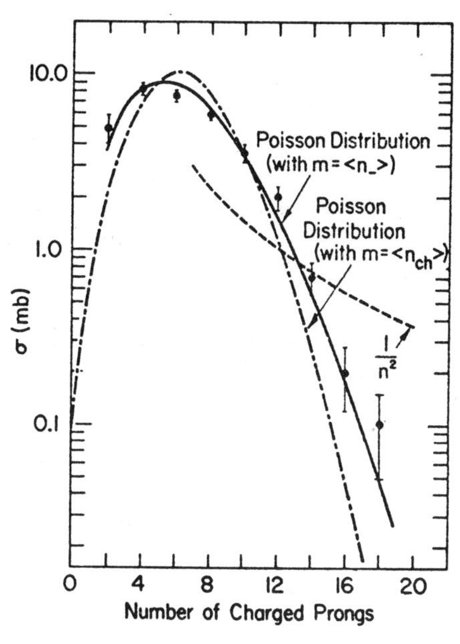

Fig. 1.5. Prong distributions of pp interactions at p_L=102 GeV/c are compared with Poisson distributions in n_{ch} and in $n_- = \frac{1}{2}n_{ch} - 1$. The agreement with the latter is presumably accidental. Taken from ref. 9.

The width of the σ_n distribution would be even larger had we used the definition (1.3). Indeed, including $\sigma_{e\ell}$, one finds that σ_2 sticks out from the generally smooth distribution, thus evidently representing a term of a special nature which changes only slowly with energy. It is therefore advantageous to separate $\sigma_{e\ell}$ from the rest, which may be thought of as a smooth distribution that shifts with energy with a common scale.

It was recently shown that the formula[11]

$$\frac{\sigma_n}{\sigma_I} \approx \langle n \rangle^{-1} \; \psi\left(\frac{n}{\langle n \rangle}\right) \tag{1.8}$$

provides a phenomenological fit to the pp data. This result, although not motivated by theoretical models, includes the previous observation about the width. It implies that each moment of the

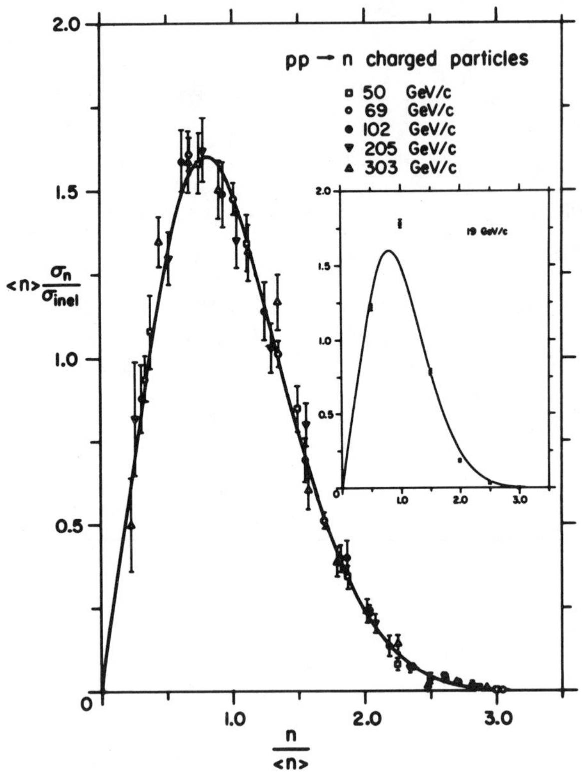

Fig. 1.6. Experimental data for $\langle n\rangle \frac{\sigma_n}{\sigma_I}$ fall on a universal curve when plotted vs. $n/\langle n\rangle$. Taken from ref. 12.

distribution obeys $\langle n^m\rangle = C_m \langle n\rangle^m$. This relation seems to be roughly true at least for the low m values. The fit to the data is shown in fig. 1.6.

To summarize this chapter we may say that the total cross sections of the hadrons obey some simple symmetries (Pomeranchuk theorems) and have very slow variations with s. The slow energy dependence is built in some tricky fashion by various increasing and decreasing components. A first glimpse of this complexity is given by the study of the topological cross sections whose gross features have been described above.

2

One-Particle Inclusive Distributions

In the previous chapter we discussed cross sections which depend only on one variable, s. The next more complicated object is inclusive production A + B → C + anything, to be denoted by (AB, C).

We define the corresponding invariant cross section by

$$\rho^{C}_{AB} = \omega \frac{d\sigma}{d^3 q}$$

$$= \frac{1}{2s} \sum_{n=2}^{\infty} \sum_{k} \int \prod_{i=1}^{n} \frac{d^3 q_i}{(2\pi)^3 2\omega_i} \omega_k \; \delta^{(3)}(q_k - q) \cdot$$

$$\cdot \delta^{(4)}(p_1 + p_2 - \sum^{n} q_j) \left| T(p_1 + p_2 \to q_1 + \ldots + q_n) \right|^2 \tag{2.1}$$

where the summation over k is carried out only over the momenta of particles of type C. The sum over n includes implicitly a sum over all possible kinds of particles and all spins. ρ_{AB}^{C} depends on three independent variables. In analogy with the two-body scattering nomenclature we may choose them as s, t, u or s, t, M^2 where

$$\begin{aligned} s &= (p_1 + p_2)^2 \equiv P^2 \\ t &= (p_1 - q)^2 \\ u &= (p_2 - q)^2 \end{aligned} \tag{2.2}$$

and $$M^2 = s + t + u - m_A{}^2 - m_B{}^2 - m_C{}^2 \tag{2.3}$$

The choice of variables to be used in the discussion depends on the physics of the problem. In the analysis of two-body scattering one uses s and t in the region of high s and low t (forward scattering) and, alternatively, s and u in the region of high s and low u (backward scattering). Since forward and backward are the regions where the cross sections are concentrated, these are evidentally convenient variables. In inclusive cross sections we will often encounter situations where all of s, t and M^2 or s, t and u are large. If the cross sections stay finite in these regions they must depend on some ratios of these variables - a phenomenon known as scaling.

In describing total cross sections we found that three rules summarized the basic experimental information. The analogous situation for inclusive distributions is more complicated because of the larger number of variables. The description of these distributions through a set of rules inevitably calls for a definite point of view since the data are always somewhat ambiguous and the simplicity and elegance of their categorization lies in the eyes of the beholder. In this chapter we define and discuss the ideas of scaling and limiting fragmentation. In a parallel discussion in chapter 6 we will present a derivation of these rules based on the language of Regge poles. The latter is at best only an approximation to the physical situation. So, also, may be some of the rules presented here. The reader should therefore be aware of the fact that small violations of these rules cannot be excluded and are perhaps even to be expected.

Rule 1: All production experiments show a strong exponential decrease in the transverse momentum ($\vec{q}_T$, perpendicular to $\vec{p}_1-\vec{p}_2$). Characteristic distributions are shown in fig. 2.1. They vary somewhat from particle to particle but all have the common characteristic of a sharp cutoff. The data in this figure can be fitted by an $\exp(-bq_T)$ form with a slope parameter b in the vicinity of 4-6 GeV^{-1}.

This fact means that we are left with only one dimension (the longitudinal) to describe all other physical effects. We may thus think of using $q_L^{c.m.}$ - the longitudinal momentum in the c.m. frame. This variable ranges from $-\frac{\sqrt{s}}{2}$ to $+\frac{\sqrt{s}}{2}$ at high energies. Since we look for phenomena that are very slowly varying we may expect that q_L has to be scaled down by $\frac{\sqrt{s}}{2}$ leading to the following

Rule 2 (hadronic scaling): The inclusive

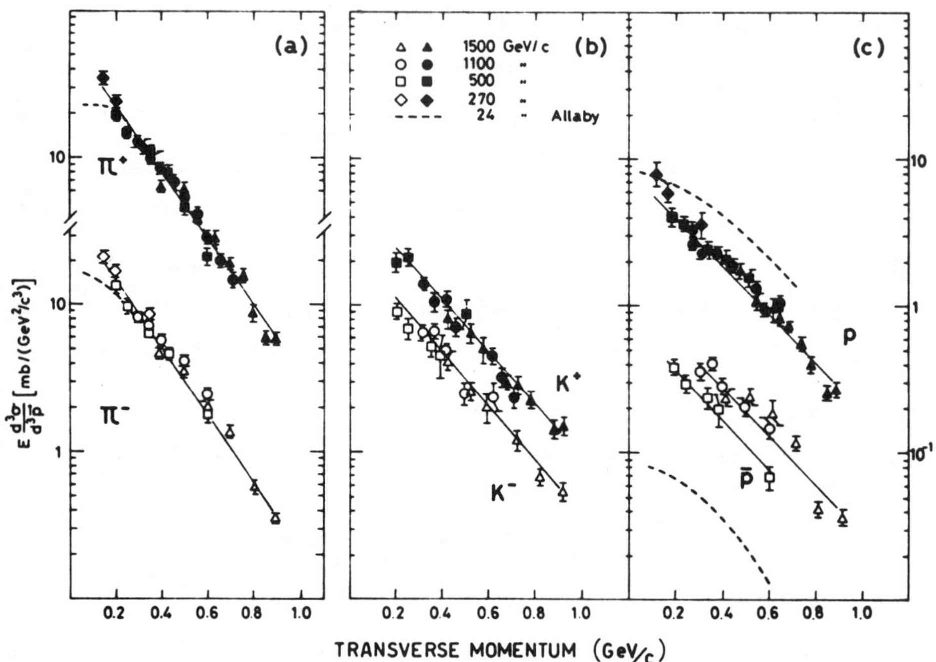

Fig. 2.1. Invariant inclusive cross sections plotted vs. q_T. ISR data are shown for various energy values (the quoted numbers represent equivalent lab momenta $\approx s/2m_N$) at x=0.16 and compared with low energy data. The straight lines represent exponential fits. Taken from ref. 13.

distributions ρ^{C}_{AB} are functions only of q_T and $x = \frac{2q_L^{c.m.}}{\sqrt{s}}$. This rule was suggested by the multiperipheral model[14] ten years ago and was recently reformulated by Feynman.[1] Experimental investigations in the past two years have shown that pion distributions obey this law quite nicely. We can see this in fig. 2.2 where a comparison is made between ISR and low energy ($s \approx 50$ GeV^2) data. The figure is impressive in view of the fact that the

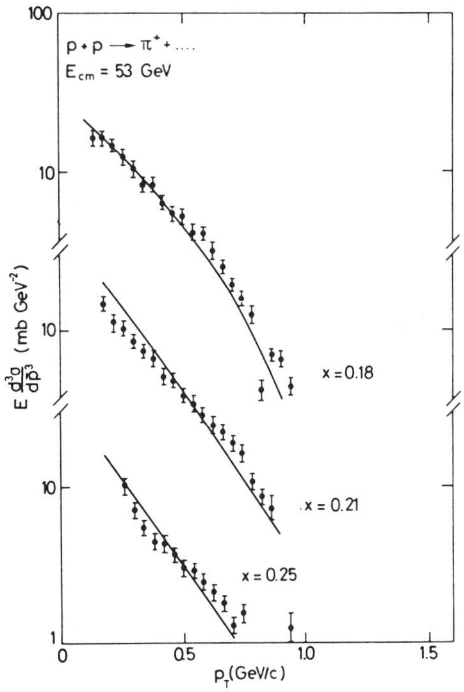

Fig. 2.2. Invariant cross sections in the ISR are compared with interpolated accelerator data at p_L=19.2 GeV. Taken from ref. 15.

total energy $\sqrt{s}$ varies by an order of magnitude. The scaling rule holds for pionic distributions over this wide range in energies. For x values very close to zero one observes some slight variations (visible in fig. 2.1a) to which we will return below. Other particles reach the scaling region at higher energies than pions. This is evident from fig. 2.1c which shows that the (pp, $\bar{p}$) data in the ISR range lie definitely higher than the 24 GeV data.

The same rule can be put in a different form in the $|x| \neq 0$ regions. We note that by boosting along the longitudinal direction we can change from the c.m. to the lab (target) frame and obtain

$$-x = \frac{2q_L^{c.m.}}{\sqrt{s}} \approx \frac{\omega^{c.m.} - q_L^{c.m.}}{E_2^{c.m.} - p_2^{c.m.}}$$

$$= \frac{\omega^{lab} - q_L^{lab}}{E_2^{lab} - p_2^{lab}} = \frac{\omega^{lab} - q_L^{lab}}{M_2} \qquad (2.4)$$

The important point here is that the final relation between x and q_L^{lab} does not involve s. Hence any function of x and $q_T^{c.m.}$ can be expressed as a function of q_T^{lab} and q_L^{lab} in the region of finite negative x and high s. Similarly one can connect the positive x region with the projectile frame. This leads then to the following

<u>Rule 3 (limiting fragmentation):</u> The inclusive distributions reach a limiting value for any specified q^{lab} as s is increased indefinitely. Such an hypothesis was suggested by BCYY[16)] who viewed the emitted particles as a result of the fragmentation of the two original particles into well-defined ratios of hadronic matter under the influence of the collision. It is interesting that this asymptotic property already holds at low energies[17)] as seen in fig. 2.3. We see that the (π^+p, π^-) data for negative q_L^{lab} momenta are essentially energy independent. On the other hand the (π^+p, π^+) data are slightly decreasing. This correction to scaling is reminiscent of transient effects in total cross sections (fig. 1.1) and can be accommodated in a Regge analysis (see chapter 6).

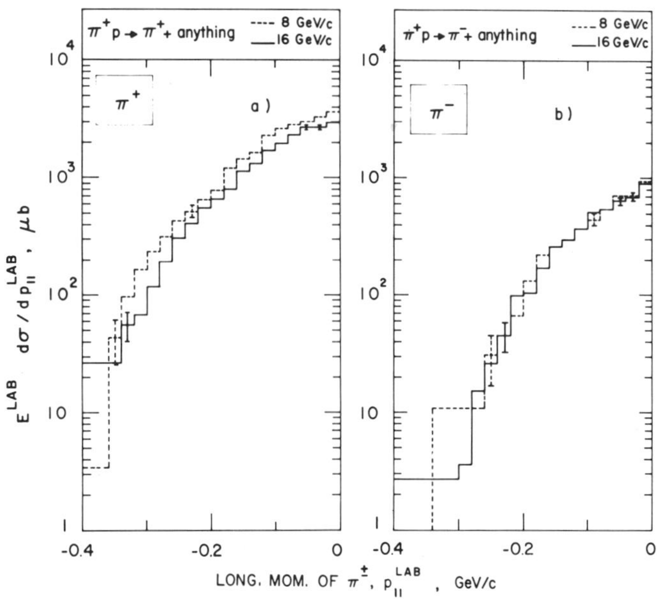

Fig. 2.3. Limiting fragmentation observed for accelerator data at lab. momenta of 8 and 16 GeV. Taken from ref. 17.

The fragmentation picture can be carried a little further by proposing that the projectile does not really affect the fragmentation products of the target; it merely acts as a catalyzer which enables the fragmentation to occur. It indeed turns out that the following rule is observed:

Rule 4: The inclusive distribution ρ_{AB}^{C} in the target fragmentation region is independent of the projectile but for an overall normalization proportional to $\sigma_T(AB)$. This rule is verified at low energies as shown in fig. 2.4. Once again some energy dependent deviations may occur and it remains to be seen how well this rule works in

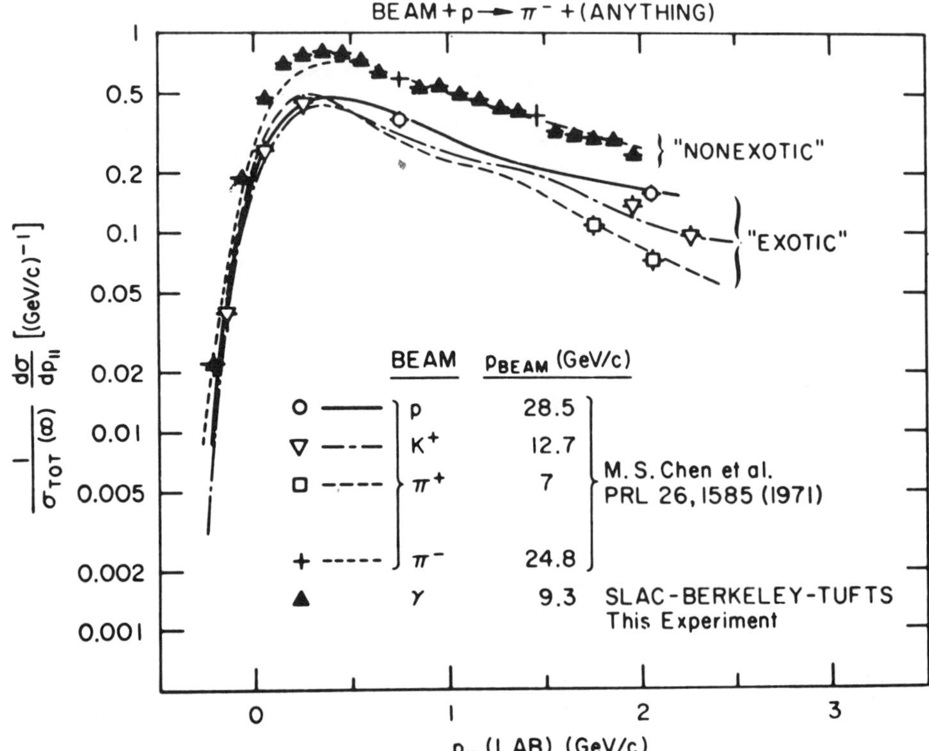

Fig. 2.4. Verification of rule 4. Taken from ref. 18.

the high energy regions of NAL.

Rule 4 can be put in a neat form by noticing that energy momentum conservation leads to the sum rule

$$\sum_C \int \frac{d^3q}{\omega}\, q^\mu\, \rho_{AB}^C = P^\mu\, \sigma_{AB} \tag{2.5}$$

which follows directly from the definition (2.1). Using the ω and q_L components of (2.5) we can write

$$\sum_C \int \frac{d^3q}{\omega}\, \frac{\omega - q_L}{M_B}\, \rho_{AB}^C = \sigma_{AB}\left(1 + \frac{\omega^A - q_L^A}{M_B}\right)$$

in the target (B) frame of reference. As s

increases $\omega^A - q_L^A \to 0$ and, using eq. (2.4), one may rewrite the sum rule in the form

$$\sum_C \int_{-1}^{0} dx\, d^2q_T \frac{1}{\sigma_{AB}} \rho_{AB}^C = 1 \qquad (2.6)$$

We have limited ourselves to the negative x region where (2.4) holds. The result (2.6) is asymptotically correct since $\omega^C - q_L^C \to 0$ for x near zero, or in the positive x region, when the momentum components are measured in the lab frame. A similar sum rule can be obtained for the projectile fragments in the positive x interval. Chou and Yang[19)] called the contribution of each kind of particle C to the sum rule (2.6) its fragmentation fraction. If rule 4 holds asymptotically it means that these fragmentation fractions are determined by particle B and are independent of particle A. To have some numbers we may quote the estimates of ref. 19 for the fragmentation fractions of a proton: 0.4 for the proton contribution, 0.12 n, 0.4 all pions, 0.05 all kaons. These estimates use $\sigma_{AB} = \sigma_T(pp)$ and low energy data. The numbers change somewhat in the higher energy regions but are still dominated by the proton and pions.

By invoking charge conservation one may obtain a charge sum rule analogous to eq. (2.5), and similarly for any other conserved additive quantum number. If, in addition, we employ the fragmentation idea, the sum rule separates into two parts, for the fragments of the projectile and of the target respectively. Thus one obtains

$$\sum_C Q_C \int_0^1 \frac{dx\, d^2q_T}{\bar{x}} \frac{1}{\sigma_{AB}} \rho^C_{AB} = Q_A$$

and
$$\sum_C Q_C \int_{-1}^0 \frac{dx\, d^2q_T}{\bar{x}} \frac{1}{\sigma_{AB}} \rho^C_{AB} = Q_B \qquad (2.7)$$

where $\bar{x}^2 = x^2 + \frac{4(m_C^{\ 2}+q_T^{\ 2})}{s}$. Such sum rules are not obeyed[20] for $s \lesssim 60\ \mathrm{GeV}^2$ and it remains to be seen whether they hold in the higher energy ranges. In particular, note that charge conservation implies that near x=0 the average charge density should vanish. We will return to this point in Rule 6.

A fragmentation picture discusses separately the target fragmentation region (negative x) and the projectile fragmentation region (positive x). The two parts are smoothly connected at x=0. The properties of inclusive distributions in the neighborhood of x=0 are of particular importance since they describe the bulk of the outgoing particles. To see this let us evaluate the sum rule

$$\int \frac{d^3q}{\omega} \rho^C_{AB} = \langle n_C \rangle \sigma_{AB} \qquad (2.8)$$

This rule follows from comparing (2.1) with (1.1). It is a consequence of the fact that each of the outgoing particles C is counted in every reaction. If we include the elastic channel in the definition (2.2) then $\sigma_{AB} = \sigma_T^{AB}$ and $\langle n_C \rangle$ is defined on the basis of all cross-sections as in eq. (1.3).

If, on the other hand, we exclude the elastic channel from the sum in the definition (2.2) then $\sigma_{AB} = \sigma_I^{AB}$ and $\langle n_C \rangle$ changes accordingly to (1.4). In any case we note that

$$\frac{d^3q}{\omega} = \frac{\pi \sqrt{s}\ q_T\ dq_T\ dx}{E} = \frac{2\pi\ q_T\ dq_T\ dx}{\sqrt{x^2 + \frac{4m_T^2}{s}}} \tag{2.9}$$

where we have used the following definition of the "transverse mass":

$$m_T^2 = m_C^2 + q_T^2 \tag{2.10}$$

Assuming Rule 2, we write

$$\rho_{AB}^C = \rho_{AB}^C\ (q_T,\ x)$$

$$\int 2\pi\ q_T\ dq_T\ \rho(q_t,\ x) = \rho(x) \tag{2.11}$$

and obtain[21)]

$$\langle n_C \rangle\ \sigma_{AB} = \int_{-1}^{1} \frac{dx\ \rho(x)}{\sqrt{x^2 + \frac{4m_T^2}{s}}}$$

$$= \rho(x=0)\ \ell n s + \text{const.} \tag{2.11}$$

We see that the important contribution to the integral comes from the central region (i.e., $x \approx 0$) and leads to a logarithmic increase of $\langle n_C \rangle \sigma_{AB}$. Thus for asymptotically constant cross-sections scaling predicts a logarithmic increase of $\langle n_C \rangle$ determined by $\rho(x=0)$.

The reason that the $x \approx 0$ region is so important comes simply from the fact that the whole

$q_L^{c.m.}$ range is scaled down by $\sqrt{s}$ and every finite $q_L^{c.m.}$ will end up at x=0 as s increases. For theoretical as well as phenomenological reasons[22,23] it is therefore advantageous to define a new variable - rapidity - whose range is proportional to ℓn s and is natural for the discussion of the central region. Let us denote rapidity by y and define it through

$$y = \ell n \frac{\omega + q_L}{m_T} = \frac{1}{2} \ell n \frac{\omega + q_L}{\omega - q_L}. \tag{2.13}$$

It follows then that

$$q_L = m_T \sinh y, \ \omega = m_T \cosh y, \ dy = \frac{dq_L}{\omega} \tag{2.14}$$

and longitudinal boosts are simple translations in rapidity $y \to y + \text{const}$. Since the energy of the outgoing particle in the c.m. is limited by $\frac{1}{2}\sqrt{s}$, the range of variation of y is $Y = \ell n(s/m_T^2)$.

In order to understand the meaning of rapidity we reproduce in fig. 2.5 a comparison between the different variables. Shown are plots in rapidity (y=0 chosen at $q_L^{lab}=0$), c.m. momentum (denoted by p_L^*), and lab momentum (p_L). The vertical axis is the transverse momentum (p_T). Comparing 2.5a and 2.5b we see that most of the rapidity range is mapped into the $x \simeq 0$ region. Comparing 2.5b and 2.5c we see that most of the negative x region comes from negative momenta in the lab (line A specifies the limit). The shaded areas correspond to the region where pions from a process $pp \to \Delta\Delta \to \pi +$ anything would be found. Note the regular form in the rapidity plot. Since longitudinal boosts lead to translations in

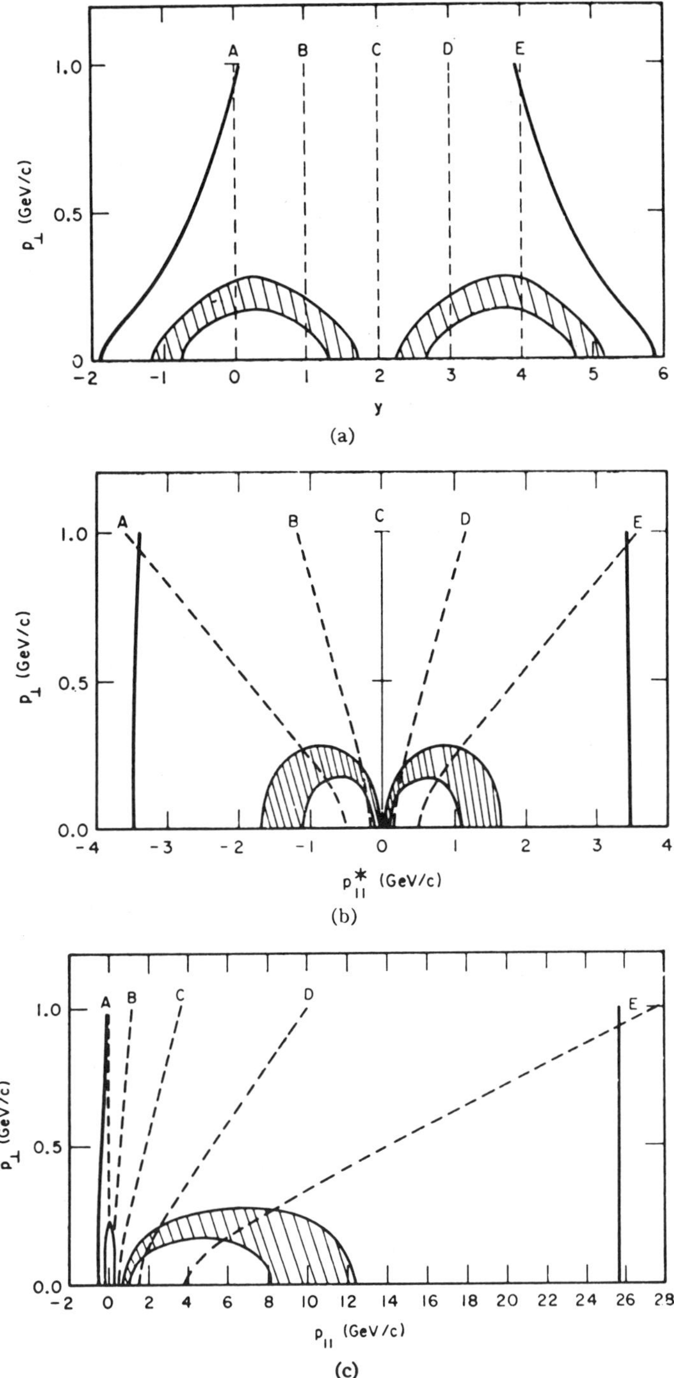

Fig. 2.5. Comparison of rapidity with c.m. and lab. momenta plots for (pp,π) at p_L=25.6 GeV. Taken from ref. 22.

rapidity one finds that plots in the latter variable conserve the symmetry that is evident in the Δ rest frame. An important property that may be seen in the plots is that for high values of $q_T >> m$ the following connection holds:

$$y \approx \eta = -\ln \frac{\tan\theta^{c.m.}}{2} \approx -\ln \tan\theta^{lab} - \ln \frac{\sqrt{s}}{2m_B} \quad (2.15)$$

This representation of y in terms of an angle is useful in experiments that can measure the angle but not the momentum. It has been widely used in cosmic ray studies[24)] and also in the recent ISR measurements of the central region.

After this digression on the meaning of rapidity let us now go back to experimental expectations. If Rule 2 is correct then one would expect to find a plateau in rapidity with the value of $\rho(q_T, x=0)$. This would then conform to (2.12):

$$\langle n_c \rangle \sigma_{AB} = \int_0^Y dy\ 2\pi\ q_T\ dq_T\ \rho(y, q_T) \approx Y\rho(x=0) \quad (2.16)$$

Y is the range of integration which, we recall, grows like $\ln s$. We expect, of course, that near the ends of this region the distributions will terminate in a fashion which reproduces the fragmentation regions of the target and the projectile. Fig. 2.6 shows the results of recent ISR experiments. They consist of 90° data (whose y values are indicated by arrows) as well as data at lower angles (less than 10°). Fig. 2.7 shows a constant behavior of $\pi^\pm$ production for a small

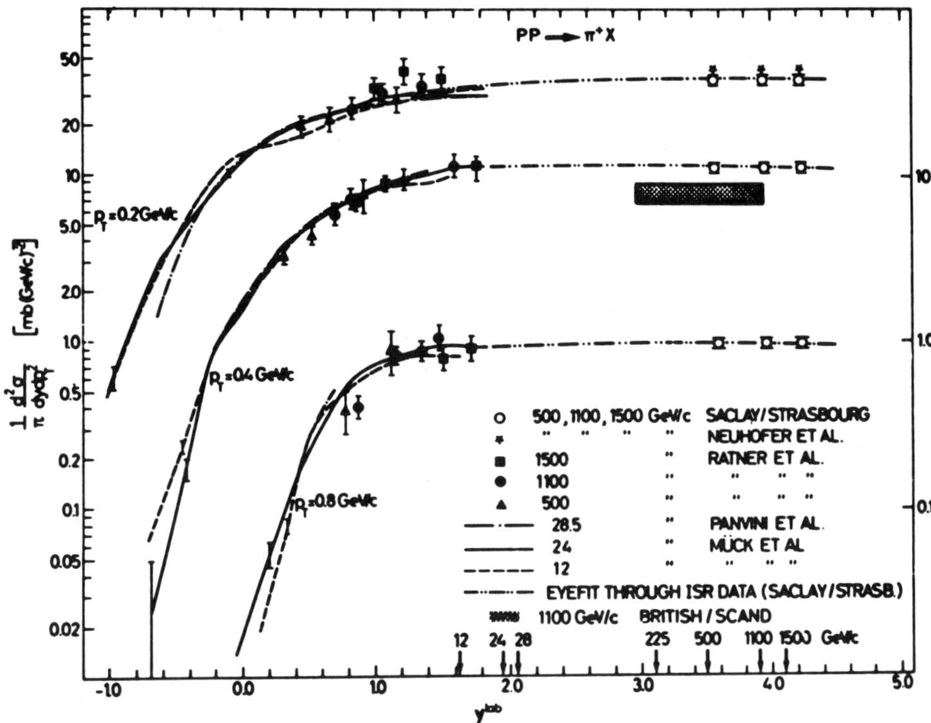

Fig. 2.6. (a) Compilation of (pp,π^+) data for different p_T values. Taken from ref. 25.

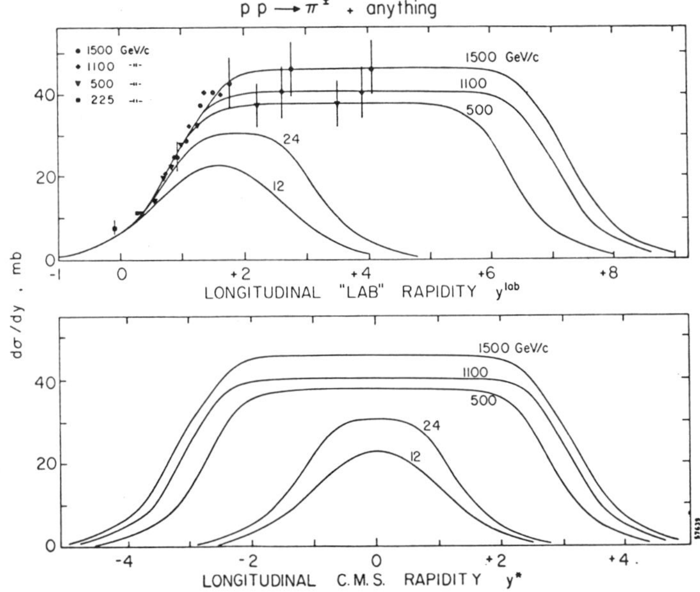

Fig. 2.6. (b) Inclusive cross sections integrated over transverse momenta. Figure taken from ref.26.

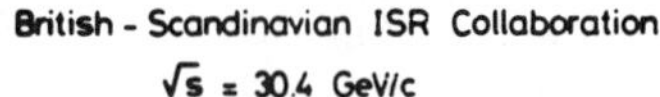

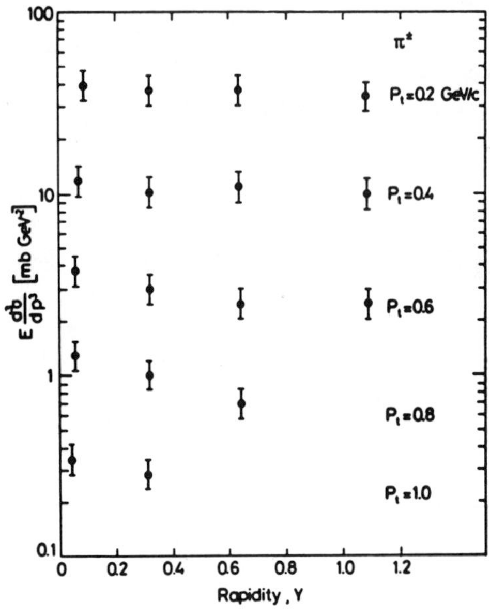

Fig. 2.7. Rapidity plateau observed for $(pp,\pi^{\pm})$ in the ISR. Taken from ref. 27.

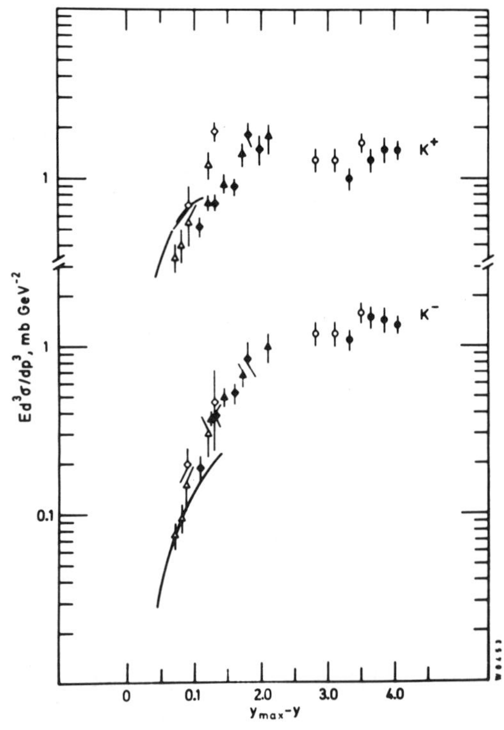

Fig. 2.8. Data for $(pp,K^{\pm})$ at $q_T=0.4$ GeV and various energies in the ISR. Taken from ref. 28.

range in c.m. rapidity. This seems to substantiate the expectations of Rule 2 and leads to

Rule 5: Inclusive distributions develop a plateau in rapidity in the central region. This together with limiting fragmentation (Rule 3) is essentially equivalent to the hadronic scaling assumption (Rule 2). The plateau region is far away from both target and projectile and may therefore be expected to display some clear symmetries. Indeed this seems to be the case;

Rule 6: There exists equal production of particles and antiparticles in the central region. This is to be expected because of overall charge, baryon number and strageness conservation, as implied by eq. (2.7). Within the ISR range one finds that these results are obeyed for both pions and kaons. In fig. 2.8 we see recent $K^{\pm}$ data that show this effect. This rule does not hold yet for p and $\bar{p}$ production in the ISR range. Although they do approach each other as s grows they did not reach yet equality in the central region.[7)] Recent observations of π^{o} production indicate that their distribution is equal to that of the other pions. These results are shown in fig. 2.9. Hence we may also make the statement that there exists an equal amount of production of each member of an isospin multiplet. This sounds very similar to our Rules 2 and 3 for total cross sections. Indeed we will see that in the Regge description of inclusive cross sections these two statements are connected - both are properties of

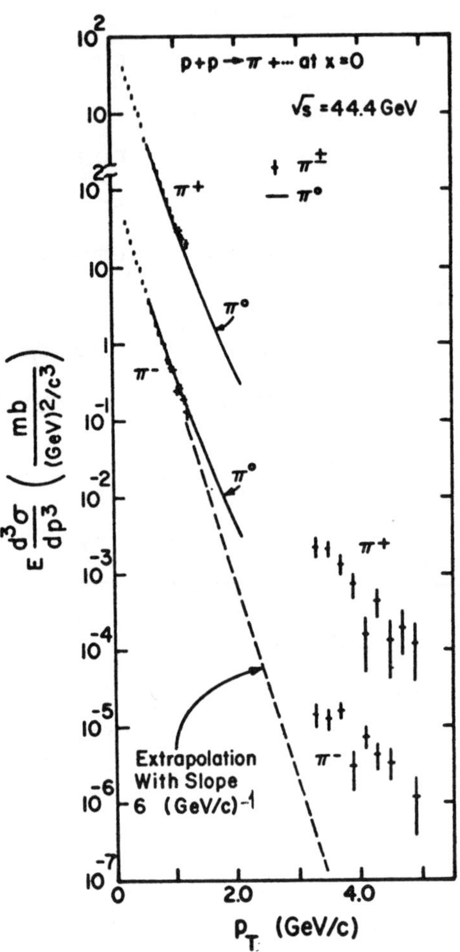

Fig. 2.9. Comparison between neutral and charged pion production at ISR. Note also the deviation from the form $\exp(-6\ p_T)$ at large transverse momentum values. Taken from ref. 29.

the Pomeron, reflecting charge conjugation and isospin invariance. One may now wonder about SU_3 invariance. Is it a possible symmetry? The answer is no; the assumption of SU_3 symmetry is self-contradictory. The argument is based on the fact that most of the known resonance decays are pionic ones (due to the symmetry breaking manifested in the inequality of the masses of the π and K mesons). If one were to produce equal amounts of all the vector mesons in the 1^- octet, one would end up with many more pions than kaons, thus breaking such a symmetry on the level of the pseudoscalar octet. It is interesting to note that also SU_2 breaking may propagate into the game by similar reasoning. Thus if there exists a nonnegligible amount of η production it will result in an observable difference between the number of π^o and the number of π^+. The results of fig. 2.9 show no such difference.

Experimental observations show a very strong dominance of pion production over K-mesons or baryon pairs,thus establishing

Rule 7: Most of the particles produced in multiparticle reactions are pions. This is already evident from the scales on fig. 2.1. The dominance of the pionic inclusive distributions is mainly observed in the central region, which is often also referred to as the pionization[30] region. A comparison between the various $\langle n \rangle$ values for different types of particles produced in pp reactions is shown in fig. 2.10.

In the fragmentation regions one finds spectra that are characteristic of the target or projectile respectively. Fig. 2.11 shows the spectrum of the proton fragmentation region at 19 GeV. There are slight changes in details as one increases the energy ; however, the general features are similar at the ISR. The p distribution is dominant for all $x \geq 0.4$. This outgoing proton is often referred to as the leading particle. At

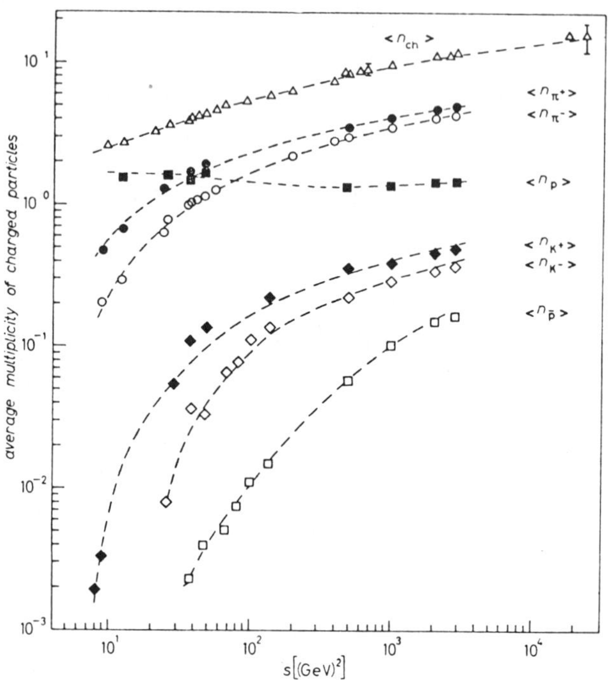

Fig. 2.10. Average multiplicities of charged particles produced in pp reactions. Taken from ref. 31.

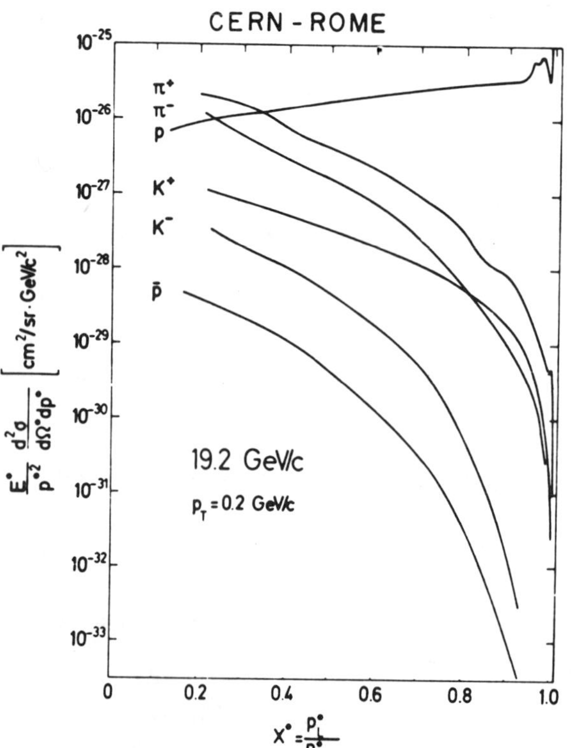

Fig. 2.11. Particle spectra in the fragmentation region. Data taken from pp reactions.[32]

low x values one can see that the π-distribution takes over. Some of the details of these distributions will be discussed in chapter 6.

Finally, we would like to note that special attention was recently given to the large transverse momentum data. As can be seen in fig. 2.9, one observes a deviation from the exponential fall-off at large q_T. An intriguing possibility is that one encounters here a power behavior of the type $p_T^{-4}\ f(\frac{t}{s}, \frac{u}{s})$ which may be expected from certain parton models.[33] This may open a new region with characteristics different from those observed in the fragmentation and central regions.

The rules on scaling, limiting fragmentation and the rapidity plateau have been presented in this fashion because of theoretical biases. Nevertheless they may be regarded phenomenologically as approximate descriptions of the experimental situation. The word approximate is of key importance in this statement. As is often the case, one can explain the same phenomena with more than one model. The various models may then differ in details that cannot be settled within present experimental accuracy, such as power or logarithmic increase of the multiplicity, exact or approximate scaling, fine structure in rapidity distributions, etc. We therefore must keep an open mind, and regard all rules with both appreciation and suspicion.

We started this chapter with a discussion of variables. Let us close it with a concise summary of the relevant variables and the relations between them. We present the formulas in the limit in which the masses of the individual particles can be neglected in comparison with s. Eq. (2.3) then leads to

$$M^2 = s + t + u \tag{2.17}$$

The transverse momentum and the scaling variables can be expressed as

$$\begin{aligned} m_T{}^2 = m_C{}^2 + q_T{}^2 &= \frac{tu}{s} + x\left[m_A{}^2(1-x)+m_C{}^2\right] \quad x>0 \\ &= \frac{tu}{s} - x\left[m_B{}^2(1+x)+m_C{}^2\right] \quad x<0 \end{aligned} \tag{2.18}$$

and $$x = \frac{t-u}{s} \tag{2.19}$$

from which it follows that

$$\frac{M^2}{s} = 1 - x + 2\frac{t}{s} = 1 + x + 2\frac{u}{s} \tag{2.20}$$

The rapidity can be obtained by writing the formula

$$x = \frac{2m_T \sinh y}{\sqrt{s}}$$

where y=0 is the c.m. position. For large values of y one can use the approximation

$$y \approx \ln\frac{x\sqrt{s}}{m_T} = \ln x + \frac{Y}{2} \qquad y>>1 \tag{2.21}$$

3

Several-Particle Inclusive Distributions and Correlations

In analogy with our definition of the single particle inclusive distribution in eq. (2.1) we may define the inclusive distribution of two particles (AB, CD) by

$$\rho_{AB}^{CD} = \omega_C \omega_D \frac{d\sigma}{d^3q_C d^3q_D} = \frac{1}{2s} \sum_{n=2}^{\infty} \sum_{k\neq \ell} \int \prod_{i=1}^{n} \cdot$$

$$\cdot \frac{d^3q_i}{(2\pi)^3 2\omega_i} \omega_k \omega_\ell \; \delta^{(3)}(q_k - q_C)\, \delta^{(3)}(q_\ell - q_D) \cdot$$

$$\cdot \delta^{(4)}\left(p_1 + p_2 - \sum_{j=1}^{n} q_j \right) |T(p_1+p_2 \to q_1 + \ldots + q_n)|^2 \tag{3.1}$$

where the summation over k and ℓ is carried out only over the momenta of particles of types C and D respectively. One may obtain a sum rule, analogous to eq. (2.8),

$$\int \frac{d^3q_C}{\omega_C} \frac{d^3q_D}{\omega_D} \rho_{AB}^{CD} = \langle n_C n_D - \delta_{CD} n_C \rangle \sigma_{AB} \ . \qquad (3.2)$$

Let us now define the correlation function[23,34)]

$$f_{AB}^{CD} = \frac{1}{\sigma_{AB}} \rho_{AB}^{CD} - \frac{1}{\sigma_{AB}} \rho_{AB}^{C} \frac{1}{\sigma_{AB}} \rho_{AB}^{D} \qquad (3.3)$$

using eq. (3.2) together with (2.8) we obtain

$$\int \frac{d^3q_C}{\omega_C} \frac{d^3q_D}{\omega_D} f_{AB}^{CD} = \langle n_C n_D - \delta_{CD} n_C \rangle - \langle n_C \rangle \langle n_D \rangle$$

$$\equiv f_2(AB,\ CD). \qquad (3.4)$$

If the production of particles were completely independent then the right hand side would vanish, indicating the absence of correlations. In the special case C=D one obtains the familiar expression $\langle n(n-1) \rangle - \langle n \rangle^2$, which vanishes for a Poisson distribution. However, totally uncorrelated production cannot, in fact, occur because of energy momentum conservation, which imposes the constraints

$$\sum_D \int \frac{d^3q}{\omega} \rho_{AB}^{D}(q)\ q^\mu = P^\mu \sigma_{AB}$$

and

$$\sum_D \int \frac{d^3q_2}{\omega_2} \rho_{AB}^{CD}\ (q_1,\ q_2)\ q_2{}^\mu = (P-q_1)^\mu \rho_{AB}^{C}(q_1)$$

$$(3.5)$$

These are direct consequences of the definitions (2.1) and (3.1). They imply

$$-\sum_{D}\int\frac{d^3q_2}{\omega_2}\, f_{AB}^{CD}(q_1,\ q_2)\ q_2{}^{\mu} = \frac{1}{\sigma_{AB}} q_1{}^{\mu} \rho_{AB}^{C}(q_1), \tag{3.6}$$

hence the left hand side cannot vanish, and hence f_2 cannot be zero.[35)] Correlations must exist.

Two particle distributions, as well as correlation functions, depend on six variables. Scaling would imply that

$$\rho_{AB}^{CD} \to \rho_{AB}^{CD}\ (x_1,\ \vec{q}_{T_1},\ x_2,\ \vec{q}_{T_2}) \quad \text{fragmentation region} \tag{3.7}$$

where we leave the vector notation on the transverse momenta since a nontrivial dependence on $\vec{q}_{T1} \cdot \vec{q}_{T2}$ may occur. When one discusses the central region with both x_1 and x_2 near zero, it is, as before, advantageous to switch to a rapidity description. Since a plateau is expected in the single particle distribution, double production can depend only on the difference between the rapidities, thus

$$\rho_{AB}^{CD} \to \rho_{AB}^{CD}\ (y_1 - y_2,\ \vec{q}_{T_1},\ \vec{q}_{T_2}). \quad \text{central region} \tag{3.8}$$

These are, of course, expectations based on assuming that the rules of the previous sections hold literally. If they are modified, then these equations will be modified too.

The effect of the conservation laws will be evident in the fragmentation regions. It is very easy to see that energy momentum conservation forbids the emission of two particles in the regions where $|x_1+x_2| > 1$. Therefore, f_{AB}^{CD} must be negative in these regions. Moreover, using (3.6) in a similar fashion to our derivation of the fragmentation sum rule (2.7), we note that asymptotically

$$(|x|\pm x)\,\frac{1}{\sigma_{AB}}\,\rho_{AB}^{C} = -\sum_{D}\int d^2q_T'dx'\cdot$$

$$\cdot f_{AB}^{CD}\,(x,\,\vec{q}_T,\,x',\,\vec{q}_T')\,\frac{|x'|\pm x'}{|x'|} \qquad (3.9)$$

which, for positive x values, results in two equations which may be written in the form

$$2\,x\,\frac{1}{\sigma_{AB}}\,\rho_{AB}^{C} = -\sum_{D}\int d^2q_T'dx'\cdot$$

$$\cdot\; f_{AB}^{CD}(x,\vec{q}_T,x',\vec{q}_T')\;2\theta(x') \qquad x>0 \quad (3.10)$$

and $$\sum_{D}\int d^2q_T'dx'\;f_{AB}^{CD}\;(x,\vec{q}_T,x',\vec{q}_T')\;\theta(-x') = 0 \quad x>0. \qquad (3.11)$$

Hence the integral over all correlation functions of two particles which are fragments of the same incident particle is negative, whereas the same quantity evaluated for fragments of different sources will vanish.

The energy momentum constraints should not be felt, in contrast, in the central region. Therefore, by looking for correlations between particles with low x values one tests more basic properties of the dynamics of multiparticle production. In analogy with the calculation that leads to (2.11) we might expect from scaling in x and the strong damping in q_T that

$$\sigma_{AB}\, C_2(AB,\, CD) \equiv \int \frac{d^3q_1}{\omega_1}\, \frac{d^3q_2}{\omega_2}\, \rho_{AB}^{CD}\, (q_1 q_2) \sim$$

$$\sim \ln^2 s \int d^2q_{T_1}\, d^2q_{T_2}\, \rho_{AB}^{CD}(x_1=0, \vec{q}_{T_1}, x_2=0, \vec{q}_{T_2}). \tag{3.12}$$

This $\ln^2 s$ behavior means, in the rapidity representation, that ρ_{AB}^{CD} behaves like a constant as $y_1 - y_2$ increases. This is the same asymptotic behavior as that of the product $\rho_{AB}^{C}\, \rho_{AB}^{D}$. The important question, then, is whether it cancels out in the correlation function f_{AB}^{CD}. Theoretical models, such as a multiperipheral or Mueller Regge pole model (see chapter 6), as well as dual models,[36] suggest that it does cancel out and that f_{AB}^{CD} depends on $(y_1 - y_2)$ as a decreasing exponential and is independent of $y_1 + y_2$. This then means that

$$\int d^2q_{T_1}\, d^2q_{T_2}\, f_{AB}^{CD}\, (q_1 q_2) = g\, e^{-\frac{|y_1 - y_2|}{\lambda}} \tag{3.13}$$

where λ is defined as the correlation length,[23] and is suggested to be of the order of 2 by the theoretical models. If this is indeed the case then

$$f_2(AB, CD) = \int \frac{d^3q_1}{\omega_1}\, \frac{d^3q_2}{\omega_1}\, f_{AB}^{CD}(q_1, q_2) = 2Y\lambda g \tag{3.14}$$

where $Y \sim \ell n s$ is the range of rapidity available. This situation is referred to as short-range correlation. If a strong damping such as that in eq. (3.13) does not hold one speaks of a long-range correlation. In particular, if f_{AB}^{CD} goes to a constant one finds $f_2 \sim \ell n^2 s$. A stronger increase of C_2 and f_2 can occur in models which violate scaling in x.

The numbers C_2 and f_2 which are defined in eqs. (3.12) and (3.14) have a directly observable meaning:

$$C_2(AB,CD) = \langle n_C n_D - n_C \delta_{CD} \rangle \tag{3.15}$$

$$f_2(AB,CD) = \langle n_C n_D - n_C \delta_{CD} \rangle - \langle n_C \rangle \langle n_D \rangle .$$

For identical particles (C=D) we see that

$$f_2 = \langle n^2 \rangle - \langle n \rangle^2 - \langle n \rangle = D^2 - \langle n \rangle \tag{3.16}$$

using the notation of (1.6). $f_2=0$ leads to the condition $D^2=\langle n \rangle$ characteristic of Poisson distributions. Thus, we learn that the question of whether the correlations are of a short-range or long-range character can be determined by the distribution of cross sections as well as by the differential form of the inclusive production of two particles. We have already seen in fig. 1.4 such results obtained for the class of all charged particles. The previous discussion holds, of course, also when C stands for "a charged particle" rather than a particle like, say, π^-. The data show that $D=1/2\langle n \rangle$, and hence, $f_2=1/4\langle n \rangle^2-\langle n \rangle$ and increases like $\ell n^2 s$ if $\langle n \rangle$ increases like $\ell n\ s$. This is evidently a sign of the existence of long-range correlations.

Experimentally, one seems to observe both of these effects. The short-range correlations cause $\frac{1}{\sigma}\frac{d^2\sigma}{dy_1 dy_2}$ to peak around $y_1 \approx y_2$, while the long-range correlations are responsible for large tails of these distributions that do not always cancel out with $\frac{1}{\sigma^2}\frac{d\sigma}{dy_1}\frac{d\sigma}{dy_2}$. As examples, let us show some recent ISR results for the ratio $R = \sigma_I \frac{d^2\sigma}{dy_1 dy_2} / \frac{d\sigma}{dy_1}\frac{d\sigma}{dy_2}$. The rapidity here is approximated by the angle, using eq. (2.15). In fig. 3.1 we show recent results for correlations between charged particles, between neutrals (γ-rays) and correlations between these two groups.

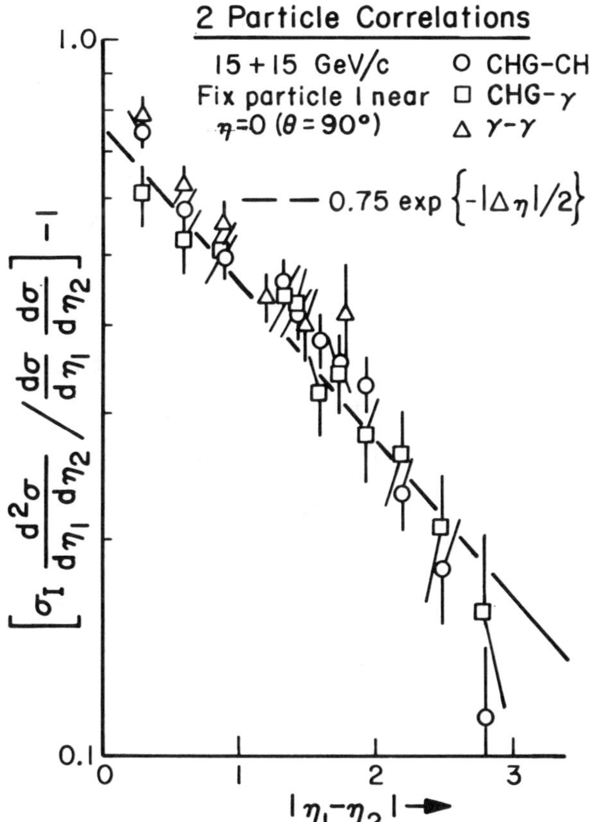

Fig. 3.1. Correlations between different types of particles show exponential decrease in the data of ref.37.

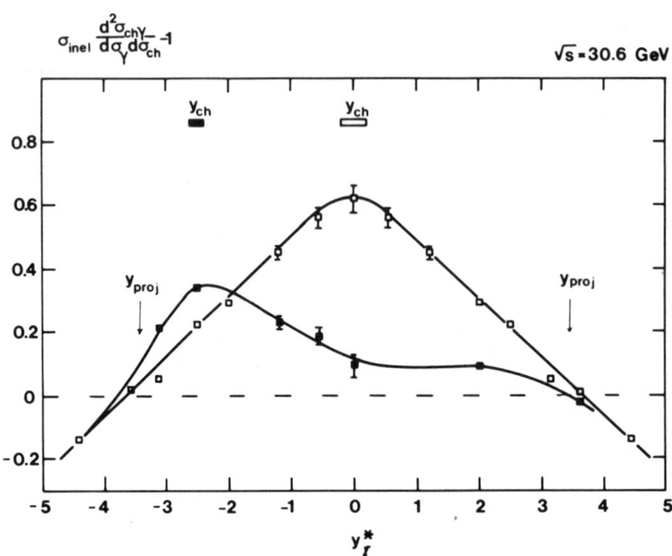

Fig. 3.2. Differential correlation ratio between γ-rays and charged particles. From ref. 38.

All seem to fall exponentially. A different experiment[38)] results in the data of fig. 3.2 which is once again the differential correlation ratio between γ-rays (mainly indicating π^0 mesons) and charged particles. We see that it peaks always around $y_\gamma = y_{ch}$; however, its functional form is different from that shown in the previous figure. Although the details of the distributions are still somewhat controversial we can be content with the fact that both show clear correlation signals which have similar qualitative features. The peak at $y_\gamma \simeq y_{ch}$ is clearly brought out in the contour diagram of fig. 3.3 which shows γ-charge correlations over the whole range of y_γ and y_{ch}.

It is important to emphasize that the same question, namely, that of the existence of short or long-range correlations, can be tested in two

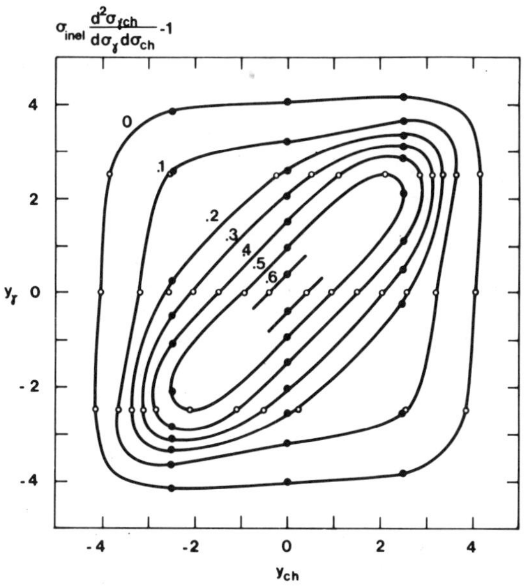

Fig. 3.3. Contour of correlation ratio values for γ-rays and charged particle differential distributions. Data of ref. 38.

different ways. One is the study of multiplicity distributions and the other is the study of the differential correlation function. The connection between the differential distributions and the correlation functions, and the definition of the latter for more than two particles, can easily be described in terms of a generating function[34)]

$$Q = \frac{1}{\sigma} \sum_{n=0}^{\infty} \sigma_n z^n \qquad \sigma = \sum_{n=0}^{\infty} \sigma_n \tag{3.17}$$

where, for simplicity, we consider only one type of particle. This can be generalized in an obvious fashion to different kinds of particles. It is now interesting to note that

$$\begin{aligned} C_N &= \frac{1}{\sigma} \sum_{n=N}^{\infty} n(n-1)(n-2)\ldots(n-N+1)\sigma_n \\ &= \frac{\partial^n}{\partial z^n} Q\Big|_{z=1} . \end{aligned} \tag{3.18}$$

This quantity however is the same as

$$C_N = \frac{1}{\sigma}\int dy_1 dy_2 \ldots dy_N \, \frac{d\sigma}{dy_1 dy_2 \ldots dy_N} \qquad (3.19)$$

and is the straight-forward generalization of C_2 defined in (3.12). We may now rewrite (3.17) in the form

$$Q = \sum_n C_n \frac{(z-1)^n}{n!} \; . \qquad (3.20)$$

It is interesting to compare these forms with a Poisson distribution:

$$\frac{\sigma_n}{\sigma} = e^{-\bar{n}} \frac{\bar{n}^n}{n!} \qquad Q_P = e^{-\bar{n}} \sum_n \frac{(\bar{n}z)^n}{n!} = e^{\bar{n}z - \bar{n}} \; . \qquad (3.21)$$

A comparison with (3.20) shows that

$$C_n = \bar{n}^n \qquad (3.22)$$

is the property that defines the Poisson distribution. An alternative statement is that in this case $\ell n\ Q(z)$ is linear in z. Since this is the classical case of no correlations we will define the general correlation coefficients by

$$f_n = \frac{\partial^n}{\partial z^n} \ell n\ Q\Big|_{z=1} \qquad n \geq 1 \qquad (3.23)$$

and hence all f_n with $n \geq 2$ vanish for a Poisson distribution. A straight-forward calculation establishes the agreement with the previous definition of f_2 in eq. (3.15). Comparing the resulting expression

$$\ell n\ Q = \sum_n f_n \frac{(z-1)^n}{n!} \tag{3.24}$$

with eq. (3.20), one can derive the general connection between f_n and C_n. It turns out to be[34)]

$$C_n = n! \sum_{n_i} \delta(n - \sum_i n_i) \prod_{i=1}^{\infty} \left(\frac{f_i}{i!}\right)^{n_i - n_{i+1}} \cdot$$

$$\cdot \frac{1}{(n_i - n_{i+1})!} \ . \tag{3.25}$$

The lowest order terms are

$$C_0=1 \quad f_0=0 \quad C_1=f_1=\langle n\rangle \quad C_2=f_1^2+f_2$$

$$C_3 = f_1^3+3f_1f_2+f_3 \tag{3.26}$$

$$C_4 = f_1^4+6f_1^2f_2+3f_2^2+4f_1f_3+f_4.$$

All these can, of course, also be written in a differential form analogous to the one introduced above for two particle distributions. Note that energy momentum conservation implies that at any fixed s there exists an upper limit $n \leq N(s)$ for the series expansion of (3.17) as well as (3.20). Nevertheless, the correlation parameters are also defined for all higher n values, since a finite series in (3.17) implies an infinite series in (3.24).

A question of major importance is, as we have already seen, the energy, or rapidity, dependence of the various quantities. Short-range correlations mean that the various differential correlation functions decrease exponentially as a function of all rapidity differences. This implies that

$$f_n(Y) = \alpha_n Y + \beta_n \tag{3.27}$$

and hence

$$Q = e^{\alpha(z)Y+\beta(z)} \qquad \text{(short-range correlations)} \tag{3.28}$$

where Y is the available rapidity range, $Y \sim \ell n\ s$. In the general case we realize that the highest power of Y that appears in $\ell n\ Q$ characterizes the strongest increase of any correlation parameter. Note that even if the n in the above equations represent clusters of particles this property of the maximal energy dependence carries over to the particles themselves.[39] This fact can be demonstrated in the following way. Suppose σ_n designates the cross-section for the production of n clusters and that each cluster has a probability w_k to decay into k particles. Construct the generating function

$$g(z) = \sum_{k=0}^{\infty} w_k z^k . \tag{3.29}$$

Then it is easy to see that in terms of particle rather than cluster cross sections

$$Q = \frac{1}{\sigma} \sum_{\substack{k \\ \text{particles}}} \sigma_k z^k = \frac{1}{\sigma} \sum_{\substack{n \\ \text{clusters}}} \sigma_n \left[g(z)\right]^n ; \tag{3.30}$$

hence $Q_{\text{particles}}(z, Y) = Q_{\text{clusters}}(g(z), Y)$. (3.31)

Therefore, although the distribution of particles is not the same as the distribution of clusters, the maximal energy increase of the correlation functions remains the same. In particular, if the clusters were to show short-range correlations so, also, would the particles.

After developing the above formalism we can now return to the study of the physical behavior. Let us begin with the question of what σ one should use in the definition of multiplicities. Suppose for the moment that we were to use the total cross-section for σ_{AB} in eq. (3.2) and (3.3); in other words, that we include in σ_2 the elastic cross section. Then we could derive

$$D^2 = f_2 + \langle n \rangle = \frac{1}{\sigma_T} \sum_n (n - \langle n \rangle)^2 \sigma_n > \frac{\sigma_{e\ell}}{\sigma_T} \langle n \rangle^2 . \tag{3.32}$$

If $\sigma_{e\ell}/\sigma_T$ approaches a constant, then $f_2 \sim \langle n \rangle^2$, thus revealing long-range correlations.[40)]

In all the discussions of multiplicities in chapter 1, however, we used explicitly the inelastic cross section (excluding $\sigma_{e\ell}$). With this definition we find by a similar calculation that

$$D^2 \gtrsim \frac{\sigma_D}{\sigma_I} <n>^2 \qquad (3.33)$$

where σ_D designates diffractive production. (We will discuss diffractive scattering in the next chapter.) By definition these are the cross sections that stay roughly constant with energy. If "roughly constant" actually means a logarithmic decrease, then we find that eq. (3.32) and (3.33) do not contradict a short-range correlation picture. This, however, means that we are very far from asymptotic behavior at present energies and a relation like $D \approx 1/2<n>$ must be changed at higher energies. This is one possibility. The other extreme is to assume that the NAL data are already in the asymptotic region. In this case, in view of the empirical results we can separate σ_I further into a diffractive and a short-range component[41])

$$\sigma_I = \sigma_D + \sigma_S \qquad (3.34)$$

where σ_D and σ_S remain comparable asymptotically. If one now assumes that the cross sections that lead to σ_S define a distribution with

$$<n_S> \sim \ln s \qquad D_S \sim \ln s \qquad (3.35)$$

whereas the analogous diffractive quantities are

$$\langle n_D \rangle \sim \text{const} \qquad D_D \sim \text{const} \tag{3.36}$$

then one can write

$$D_I^2 = \frac{\sigma_S}{\sigma_I} (D_S^2 + \langle n \rangle_S^2) + \frac{\sigma_D}{\sigma_I} (D_D^2 + \langle n \rangle_D^2) - \langle n_I \rangle^2 . \tag{3.37}$$

Assuming now the empirical relation $D_I = 1/2 \langle n \rangle_I$ to represent quantities that grow like $\ell n\ s$, we can substitute it into (3.37) and equate the first and second leading orders in a $\ell n\ s$ expansion to obtain

$$\frac{\sigma_S}{\sigma_I} = 0.8 \qquad \frac{\sigma_D}{\sigma_I} = 0.2$$

$$D_S^2 = \frac{1}{2} \langle n \rangle_S \langle n \rangle_D . \tag{3.38}$$

Similar results have been obtained in many recent detailed fits.[39,42] We see that small $\langle n \rangle_D$ values are consistent with quite narrow S-distributions.

The main assumption on which this derivation hinges is that the asymptotic $\ell n\ s$ expansion correctly represents the orders of magnitude of quantities measured at available energies. The uncertainty about the exact asymptotic behavior is unfortunately unavoidable and is responsible for the simultaneous viability of different and contradictory models.

4

Diffraction in Two-Body Scattering

Elastic scattering is a diffractive process. By that we mean

Rule 1: $(d\sigma/dt)_{e\ell}$ has an energy independent, or only weakly energy dependent, structure. This is most strikingly brought out by the recent measurements of pp elastic scattering at the ISR. The data show only very slight changes over large ranges of s. A characteristic plot of $d\sigma/dt$ is shown in fig. 4.1. The energy variation is plotted in fig. 4.2 which shows $d\sigma/dt$ for fixed t values as a function of s. The ISR points display little energy dependence, whereas lower energy data show more significant changes. For comparison we show in fig. 4.3 the data below 24 GeV. Note that they lie mostly above the line $G^4(t)$ (where G(t) is the electric form factor of the proton) whereas the high energy data in fig. 4.1 lie well below it. This line appears in both figures because of previous theoretical suggestions

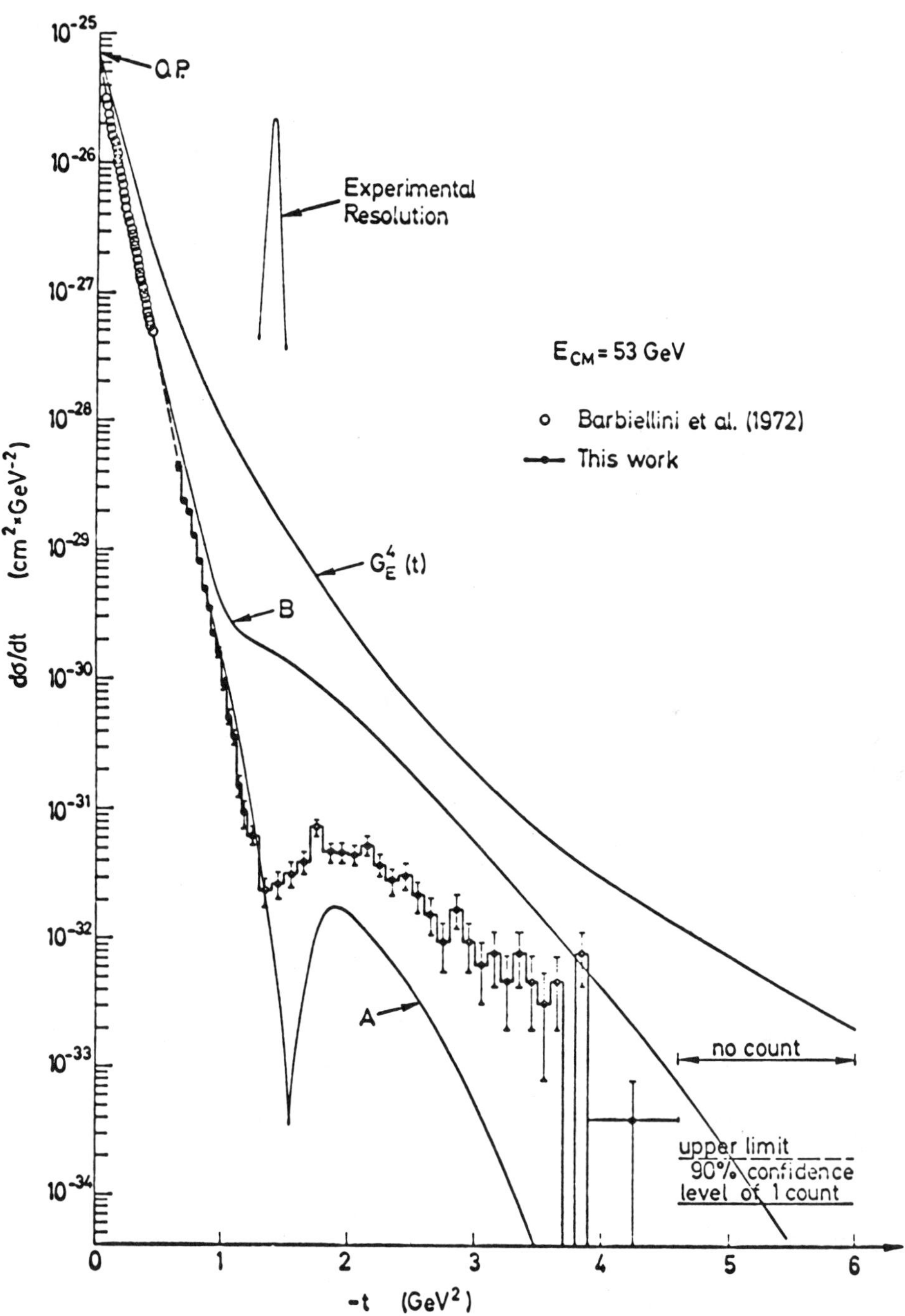

Fig. 4.1. pp elastic scattering. Taken from ref. 43.

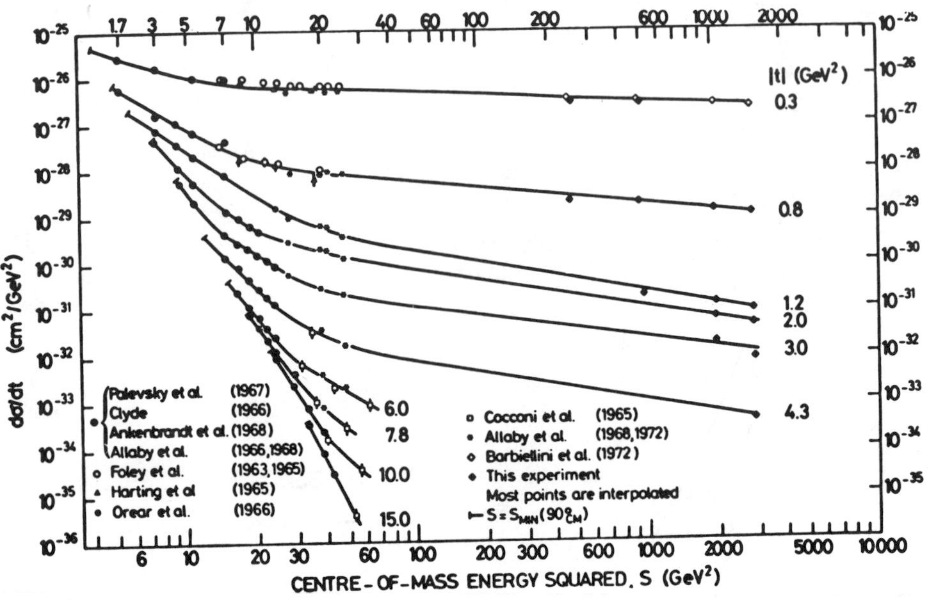

Fig. 4.2. pp elastic scattering. Taken from ref. 43.

that it might serve as a good description of the data. (See discussion in chapter 16.)

In fig. 4.3 one can clearly see a strong shrinkage with increasing energy of the pp diffraction peak. At higher energies the rate of shrinkage decreases markedly, as is evident in fig. 4.2, and it may, in fact, disappear entirely at larger values of t. For very small t, however, in the range $|t|\lesssim 0.1\ \text{GeV}^2$, some shrinkage seems to persist even throughout the ISR energy range. The data shown in fig. 4.4 suggest

Rule 2: The pp diffraction peak shrinks at a decreasing rate as energy increases. This shrinkage disappears altogether at the larger t values. The "slope parameter" b(s) plotted in fig. 4.4 is defined through the equation

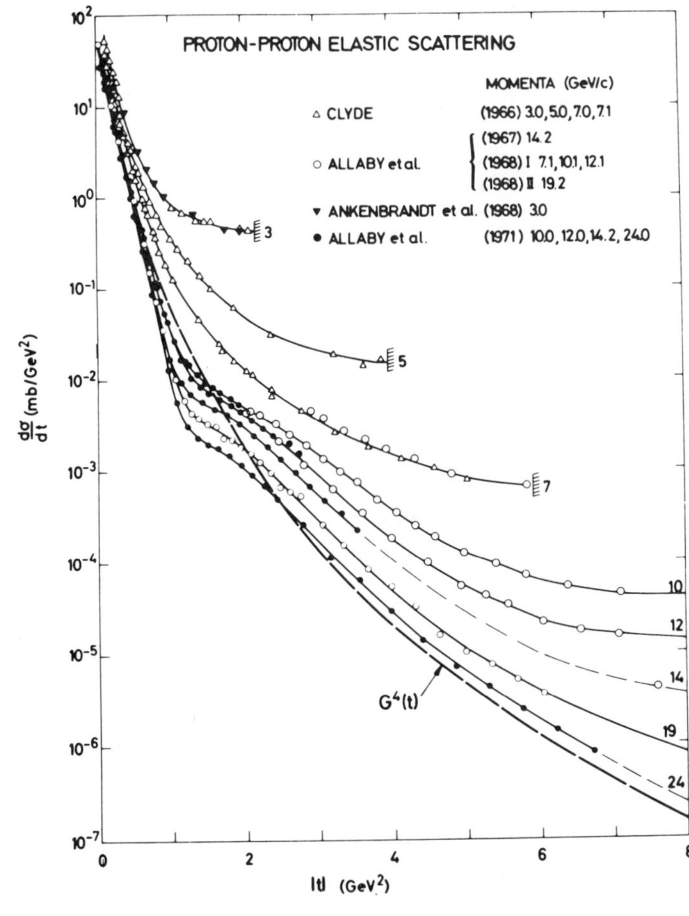

Fig. 4.3. pp elastic scattering data at accelerator energies below 24 GeV. Taken from ref. 44.

$$b(s, t) = \frac{\partial}{\partial t} \ln \frac{d\sigma/dt}{(d\sigma/dt)_{t=0}} . \tag{4.1}$$

If the diffraction peak is exactly exponential in t, b will depend only on s. Small deviations from an exponential lead to a weak t dependence in b. From the figure we see that b decreases slightly with increasing t; we also note that in the small t range ($|t| \lesssim 0.1$ GeV^2) for s up to about 150 GeV^2 b(s) can be parameterized as

$$b(s) = b_0 + 2\alpha' \ln s . \tag{4.2}$$

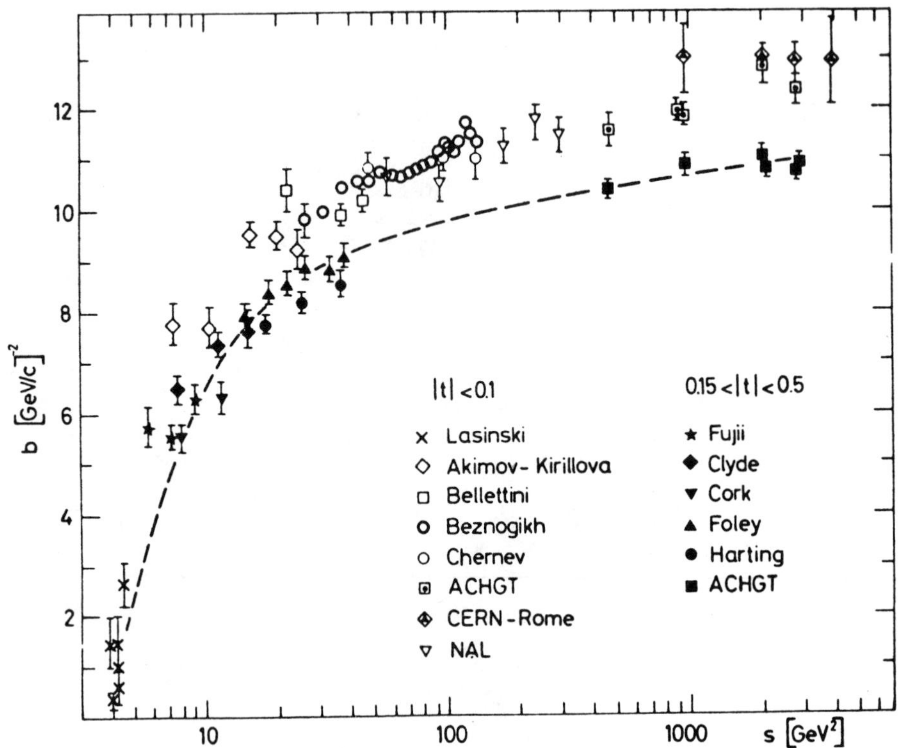

Fig. 4.4. Slope parameter in pp scattering for different t regions. Taken from ref. 45.

Conceivably we see in the ISR a roughly constant structure below $t \approx -0.1$ GeV^2 and a varying structure between $0 < t \lesssim -0.1$ GeV^2 which is also responsible for an increase in $\sigma_{e\ell}$ and σ_T (see fig. 1.2).

Elastic cross sections share many properties of the total cross sections. Thus they obey the familiar "Pomeranchuck Theorem," namely

Rule 3: $(d\sigma/dt)e\ell$ are roughly the same for particle and anti-particle cross sections. Examples of comparison between elastic cross sections are shown in fig. 4.5. Similarly one finds

Rule 4: $(d\sigma/dt)e\ell$ are about equal for members of the same isospin multiplet. Hence we may once

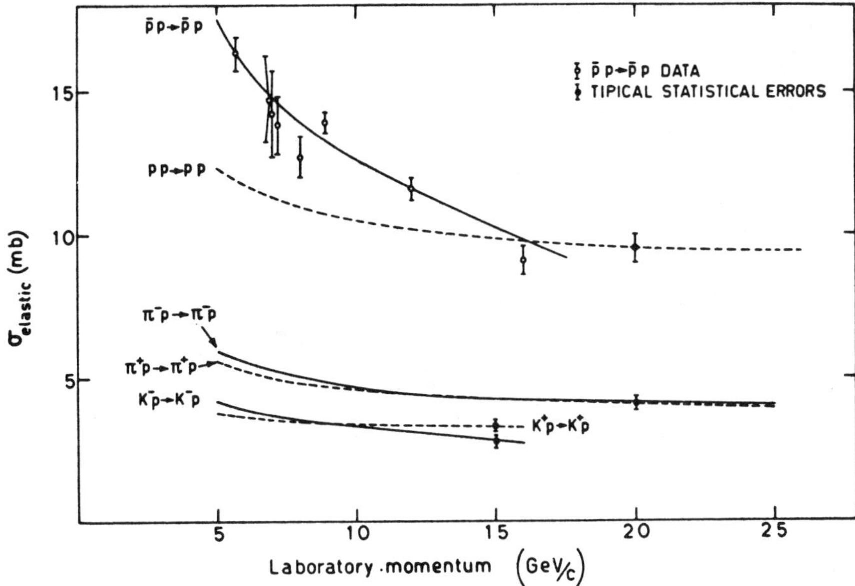

Fig. 4.5. Comparison between elastic cross sections of particles and antiparticles. Figure taken from ref. 46

again think of them as described by Pomeron exchange where the Pomeron has vacuum quantum numbers.

The connection between total cross sections and the elastic scattering amplitude is, as we have mentioned previously, established at t=0 by the optical theorem:

$$A(s,\ t=0) \equiv \text{Im}\ T_{e\ell}\ (s,\ t=0) = s\sigma_T(s). \quad (4.3)$$

By comparison of σ_T and $d\sigma/dt$ one can conclude that

Rule 5: $T_{e\ell}(s,\ t \approx 0)$ is mainly imaginary. A plot

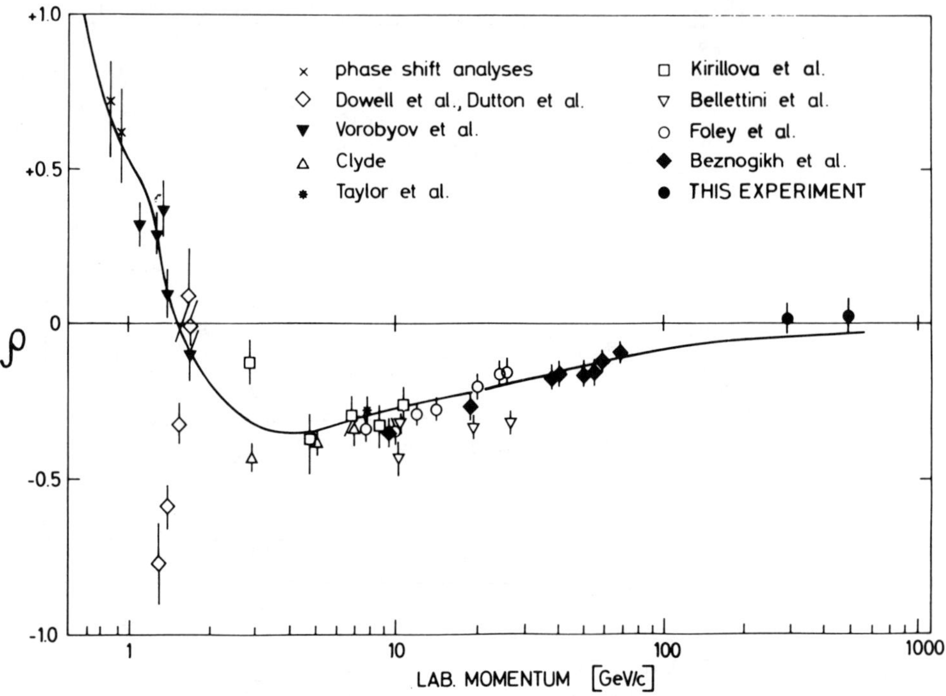

Fig. 4.6. $\rho = \mathrm{ReT}(s,t=0)/\mathrm{ImT}(s,t=0)$ for pp scattering. Taken from ref. 47.

of ReT/ImT is shown in fig. 4.6 where pp scattering data are compared with dispersion relation predictions. The small real parts are intimately connected with the equalities of the total cross sections for particles and antiparticles (see appendix C).

The fact that the pp data fit the dispersion relation well has another significance. It should be noted that the amplitude in (4.1) is the non-spin flip (or s-channel helicity conserving) amplitude. In the case of pp scattering a double s-channel flip is in principle allowed at $t \approx 0$; however, experimental data imply that it is very small. Moreover, examination of πp data and $\gamma p \to \rho p$ production suggests that elastic amplitudes are

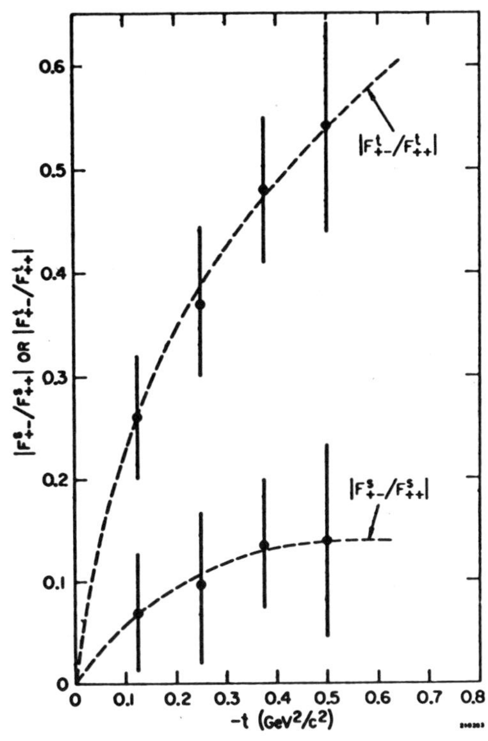

Fig. 4.7. Ratios of I=0 exchange helicity flip to nonflip amplitudes in both s-channel and t-channel c.m. frames for $\pi N \to \pi N$ at $p_L = 6$ GeV/c. Taken from ref. 49.

s-channel helicity conserving.[48] Fig. 4.7 indicates that this holds over a range of $0 < -t < 0.6$ GeV^2.

We saw the elastic scattering is largely determined by an imaginary amplitude which, in turn, can be calculated from the unitarity sum over all possible production channels. It seems quite plausible that a different quasi two-body state will have transition amplitudes into many of the same production channels. Therefore, we may also expect to find a nearly flat cross section for the production of such a quasi two-body state. This suggests the existence of a phenomenon called diffraction dissociation. The data in fig. 4.8 show that several nucleon resonances have produc-

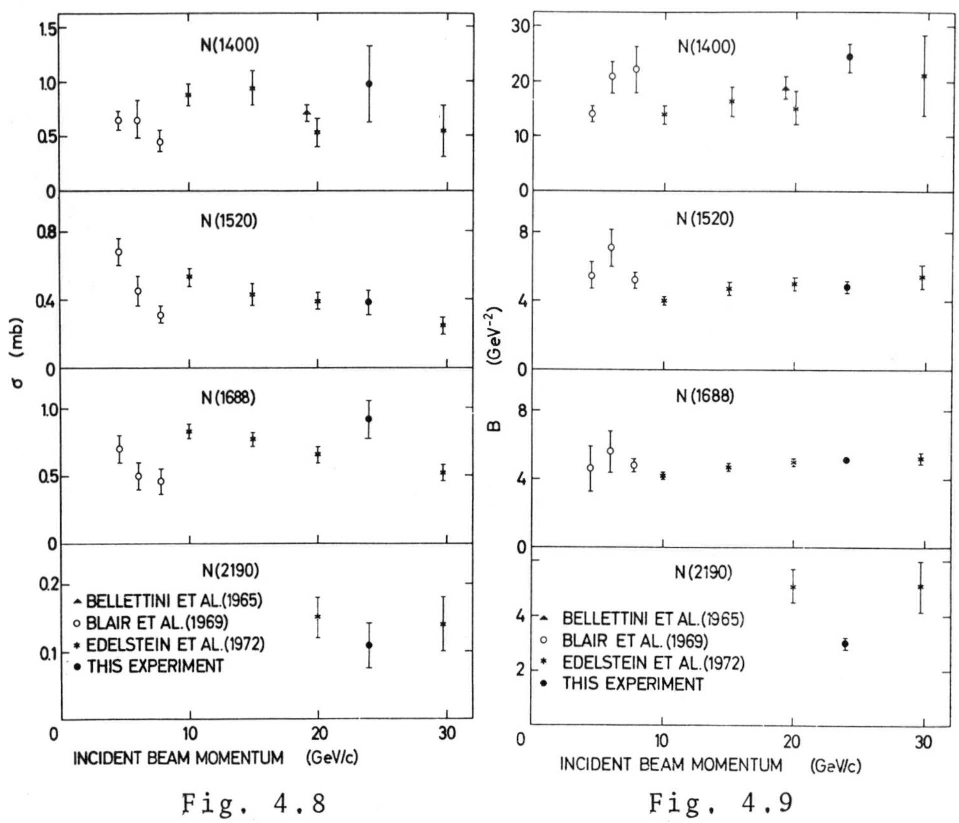

Fig. 4.8 Fig. 4.9

Figures taken from ref. 50.

tion cross sections that stay roughly the same over the whole measured energy range. The magnitude of the cross sections for the production of the individual resonances are very small; nevertheless, they clearly play a special role in the production mechanism since their cross sections are in such striking contrast to the falling cross sections of other production processes. The differential cross sections for diffraction dissociation are of the form $d\sigma/dt \sim e^{Bt}$ as in the elastic cross-sections. Data points for the slope parameters B are shown in fig. 4.9. We see that, but

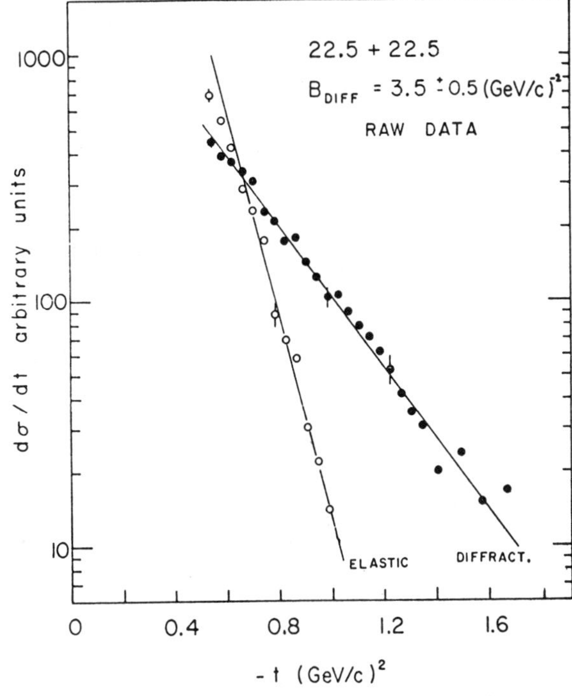

Fig. 4.10. Slope parameters for elastic scattering and diffractive dissociation at the ISR. Data of ref. 43. Compare also with fig. 6.14.

for the $N^*(1400)$, all other conspicuous resonances have slope parameters that are characteristically smaller than those for elastic scattering. This phenomenon is also observed at the ISR, as can be seen in fig. 4.10. The diffractive part in these experiments is defined as the sum over effective N* mass in $pp \to pN^*$ in the range 1.2-2 GeV.

Assuming that diffractive dissociation follows from the same basic mechanism as elastic scattering we realize that it opens up possibilities not present in the restricted set of elastic amplitudes. In particular we should mention the factorization property. Viewing diffraction as Pomeron exchange, and regarding the latter as a simple Regge pole, it was long ago suggested[51] that t-channel factorization should result in relations of the type

$$\sigma_T(AB):\sigma_T(AC) = \sigma_T(DB):\sigma_T(DC) \tag{4.4}$$

and similar relations for $d\sigma/dt$. This would imply, for instance, that

$$\left[\sigma_T(\pi N)\right]^2 = \sigma_T(NN)\ \sigma_T(\pi\pi). \tag{4.5}$$

However, $\sigma_T(\pi\pi)$ can be deduced from experiment only in a tricky way, using π-exchange models for π-production reactions. Eq. (4.5) predicts a small $\sigma_T(\pi\pi)$ of the order of 10 mb and we must await NAL experiments for a serious test of this. In the meantime, however, we can use diffraction dissociation to test predictions of the type[52)]

$$\frac{d\sigma}{dt}\ (\pi N\to\pi N^*):\ \frac{d\sigma}{dt}\ (NN\to NN^*) \approx \sigma_T(\pi N):\sigma_T(NN). \tag{4.6}$$

Data in the energy region $s<40$ GeV^2 show that the 1520 and 1688 enhancements obey this relation within experimental accuracy.

It is difficult to say whether the enhancements discussed above are single resonances or groups of resonances. In most experimental analyses a "resonance" is defined as a certain mass region. Only recently has it become possible to study diffractive production in more detail. It turns out that most of the mesonic enhancements, which were previously believed to be diffractively produced, now show some decrease with energy.[53)] Thus, for instance, the $K^+p \to Q^+p$ cross-section may be fitted by a power law P_{lab}^{-n} with $n=0.6 \pm 0.05$. The data, for different mass cuts, are shown in fig. 4.11. Once again we are left with a puzzle: Are we observing a transient falloff that will eventually disappear or will the cross sections continue to drop. Another important result from recent experiments is that, except for vector-meson photoproduction, none of the tested diffraction dissociations processes conserve s-channel helicity. It seems therefore that we are left somewhat in the dark with respect to the exact nature of diffractive production of resonances.

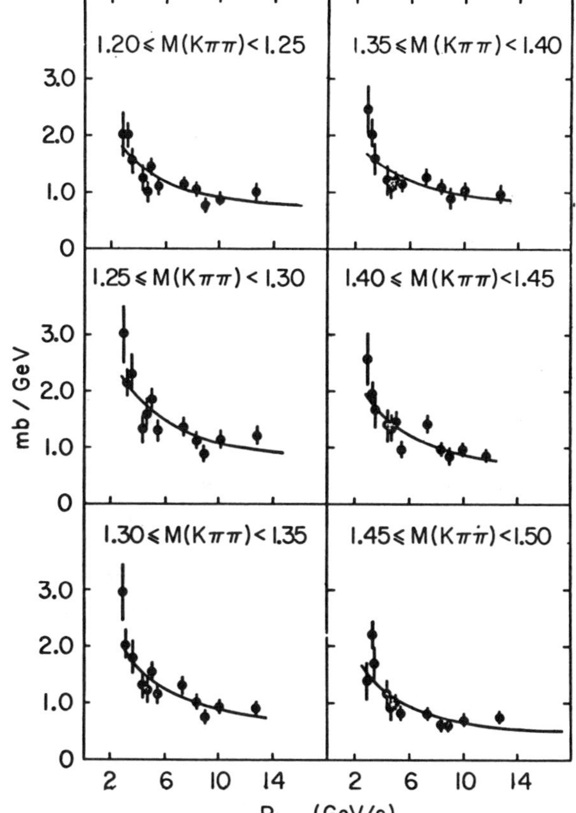

Fig. 4.11. Cross-section for $K^+p \to Q^+p$ as a function of energy for six regions of ($K\pi\pi$) mass. Taken from ref. 54.

REFERENCES

1. R. P. Feynman, Phys. Rev. Letters 23, 1415 (1969); "High Energy Collisions," Eds. C. N. Yang et al. (Gordon and Breach, New York, 1969) p.237.
2. V. Barger and D. Cline, "Phenomenological Theories of High Energy Scattering," (Benjamin, New York 1969).
3. P.G.O. Freund, Phys. Rev. Letters 20, 235 (1968). H. Harari, Phys. Rev. Letters 20, 1395 (1968).
4. R. Dolen, D. Horn and C. Schmid, Phys. Rev. 166, 1768 (1968).
5. S.P. Denisov et al., Phys. Letters 36B, 415 (1971).
6. U. Amaldi et al., CERN-Rome collaboration, Phys. Letters, B44, 112 (1973).
7. M. Jacob, Rapporteur talk at Chicago Conf. 1972, Vol. 3, p. 373.
8. F.T. Dao et al., Phys. Rev. Letters 29, 1627 (1972).
9. J.W. Chapman et al., Phys. Rev. Letters 29, 1686 (1972).
10. A.K. Wroblewski, Rapporteur's talk at the Kiev Conference (1970).
11. Z. Koba, H.B. Nielsen and P. Olesen, Nucl. Phys. B40, 317 (1972).
12. P. Slattery, Phys. Rev. Letters 29, 1625 (1972)
13. A. Bertin et al., Phys. Letters 42B, 493 (1972)
14. L. Bertocchi, S. Fubini and M. Tonin, Nuovo Cim. 25, 626 (1962).
    D. Amati, S. Fubini and A. Stanghellini, Nuovo Cim. 26, 896 (1962).
    S. Fubini, "Strong Interactions and High Energy Physics," Scottish Universities' Summer School (1963), ed. R.G. Moorhouse (Oliver and Boyd, 1964), p.259.
15. M.G. Albrow et al., Phys. Letters 42B, 279 (1972).
16. J. Benecke, T.T. Chou, C.N. Yang and E. Yen, Phys. Rev. 188, 2159 (1969).
17. J.V. Beaupré et al., Phys. Letters 37B, 432 (1971).

18. K. C. Moffeit et al., Phys. Rev. D5, 1603 (1972).
19. T.T. Chou and C.N. Yang, Phys. Rev. Letters 25, 1072 (1970).
20. D.R.O. Morrison, review talk, Oxford Conf. 1972.
21. N.F. Bali, L.S. Brown, R.D. Peccei and A. Pignotti, Phys. Rev. Letters 25, 557 (1970).
22. C.E. DeTar, Phys. Rev. D3, 128 (1971).
23. K.G. Wilson, Cornell Preprint CLNS-131 (1971).
24. L. von Lindern, Nuovo Cim. 5, 491 (1957).
25. J.C. Sens, review talk, Oxford Conf. 1972.
26. W. Kittel, CERN/D.Ph. II/Phys. 72-49.
27. B. Alper et al., paper #900, submitted to Chicago Conf. 1972. See also ref. 7.
28. E. Lillethun, proceedings of the Chicago Conf. 1972, Vol. 1, p.211.
29. M. Banner et al., cont. #478 to the Chicago Conf. (1972), Vol. 2, p.344.
30. Yash Pal and B. Peters, K. Danske Vidensk. Selsk. Mat.-Fys. Med. 33, 15 (1964).
31. M. Antinucci et al., Lett. Nuovo Cim. 6, 121 (1973).
32. J.V. Allaby et al., CERN-Rome collaboration, submitted to Oxford Conf. 1972. See also ref. 20.
33. S.M. Berman, J.D. Bjorken and J.B. Kogut, Phys. Rev. D4, 3388 (1971).
34. A.H. Mueller, Phys. Rev. D4, 150 (1971).
35. E. Predazzi and G. Veneziano, Nuovo Cim. Letters 2, 749 (1971).
L.S. Brown, Phys. Rev. D5, 748 (1972).
36. C.L. Jen, K.Kang, P. Sen and C.I. Tan, Phys. Rev. Letters 27, 458 (1971).
37. Pisa-Stony Brook collaboration, unpublished data.
38. H. Dibon et al., CERN-Hamburg-Vienna collaboration, Phys. Letters, to be published.
39. W.R. Frazer, R.D. Peccei, S.S. Pinski and C.I. Tan, preprint UCSD 10p10-113 (1972).
40. M. Le Bellac, Phys. Letters 37B, 413 (1971).
41. K. Wilson, Acta Phys. Austr. 17, 37 (1963).
A. Bialas, K. Fialkowski and K. Zalewski, Nucl. Phys. B48, 237 (1972).
42. H. Harari and E. Rabinovici, Phys. Letters 43B, 49 (1973).

L. Van Hove, Phys. Letters 43B, 65 (1973).
J.D. Jackson and C. Quigg, NAL preprint.

43. Aachen-CERN-Geneva-Harvard-Turin collaboration. Data presented at Chicago Conf. 1972. See also ref. 45.
44. J.V. Allaby et al., Phys. Letters 34B, 431 (1971).
45. G. Giacomelli, Rapporteur talk at Chicago Conf. 1972, Vol. 3, p.219.
46. G. Giacomelli, Rapporteur talk at Amsterdam Conf. 1971, p.1.
47. U. Amaldi et al., Phys. Letters 43B, 231 (1973
48. F.J. Gilman et al., Phys. Letters 31B, 387 (1970).
49. F.J. Gilman, "Experimental Meson Spectroscopy" (3rd Phil. Conf.) ed. A.H. Rosenfeld and K.W. Lai 1972, p.460.
50. J.V. Allaby et al., Nucl. Phys. B52, 316 (1973
51. M. Gell-Mann, Phys. Rev. Letters 8, 263 (1962) V.N. Gribov and I. Ya. Pomeranchuk, Phys. Rev. Letters 8, 343 (1962).
52. P.G.O. Freund, Phys. Rev. Letters 21, 1375 (1968).
53. R.E. Diebold, Rapporteur talk at Chicago Conf. 1972, Vol. 3, p.1.
54. H.H. Bingham et al., paper #218 cont. to the Chicago Conf. 1972. See report by D.W.G.S. Leith, Chicago Conf. 1972, Vol. 3, p.321.

PART II

REGGE POLE ANALYSIS

5

The Pomeron Pole and Total Cross Sections

As a first step theoretically, let us try to exploit the Regge phenomenology we have become familiar with for two-body reactions. Many good reviews,[1)] including up-to-date ones, exist of this, so here we need only sketch the two-body Regge picture in a few words.

For quantum number exchange processes at fairly low energies (s up to about 50 GeV^2) the experimental situation can be reasonably well described in terms of a few dominant Regge trajectories in each channel, with, in some cases at least, associated Regge cuts.

The characteristic features of this Regge pole picture are as follows:

1) Power dependence of cross sections:

$$\left.\frac{d\sigma}{dt}(AB \to CD)\right|_{t=0} \sim s^{2\alpha(o)-2}$$

where α is characteristic of the exchanged channel, and does not depend on the specific particles A, B, C, and D.

2) Shrinkage of forward peaks:

$$\ln\left(\frac{d\sigma/dt}{d\sigma/dt|_0}\right) \sim \alpha' t \ln s .$$

3) The correlation of the trajectory $\alpha(t) = \alpha(0)+\alpha' t$, where t is positive, with sequences of experimentally observed resonances in the relevant channel.

More detailed features of the experiments lead one into a morass of argument about just what cuts, if any, are needed; about zeros in residue functions; about exchange degeneracy of residues; and much more. In detail, the picture is muddy - but if one only looks at the broad outline, as in (1) - (3) above, the picture is rather attractive.

In this part of the book we will discuss the application of the simple Regge pole picture to diffractive and inclusive production processes. We assume, therefore, the existence in the vacuum channel of a Regge pole $\alpha(t)$, called the Pomeron, which is not dissimilar to other known trajectories occurring in channels having nonzero quantum numbers. As we shall see, many of our results will depend on the assumption that the Pomeron is a pole. Yet in reality it may well happen that the Pomeron, namely, the leading j-plane singularity structure in the vacuum channel, is not a pure pole but either is a pole with a complicated j-plane cut structure associated with it or is not a pole at all but is one or several branch points. We shall, in fact, devote much attention later to these possibilities.

If the Pomeron is, or has associated with it, j-plane cuts then many of the important properties which we shall derive in this part, such as scaling, limiting fragmentation, and short-range correlations are in principle destroyed. But experimentally, these features are at least approximately true. Hence we learn that it is worthwhile to investigate all of the results that stem from the simple Pomeron pole picture, and to try to modify them later with corrections due to cuts.

Let us accept, then, the existence of a Pomeron pole. As we shall see below, in order to accommodate experiment it is necessary to attach positive signature to this trajectory and to assume it passes through j=1 at t=0. Thus its contribution to the amplitude for any process in which the vacuum channel can be exchanged - and in particular to any elastic scattering amplitude - is, at large s and fixed t,

$$T_{AB\to AB}(s,t) = \beta_A(t)\ \beta_B(t)\ s^{\alpha(t)}\ \cdot$$

$$\cdot\left(-\frac{1+e^{-i\pi\alpha(t)}}{\sin\pi\alpha(t)}\right) \tag{5.1}$$

with $\alpha(o)=1$. The factorization of the residue, characteristic of all Regge poles, is explicitly indicated. A simple redefinition of the residue also allows the alternative form

$$T_{AB\to AB}(s,t) = \gamma_A(t)\ \gamma_B(t)\ s^{\alpha(t)}(-e^{-i\pi\alpha(t)/2}) \tag{5.2}$$

which it is sometimes more convenient to use.

From the optical theorem we can write the total cross-section for A on B as

$$\sigma_{AB}(s) = \frac{1}{s} A_{AB\to AB}\ (s,\ t=0), \tag{5.3}$$

in terms of the absorptive part A(s,t) of forward A+B→A+B elastic scattering. From eq. (5.1), we see that our assumed Pomeron pole predicts

$$\sigma_{AB}(s) = \beta_A(o)\ \beta_B(o) \equiv \beta_A\beta_B. \tag{5.4}$$

The necessary ingredients in this prediction are the intercept of one ($\alpha(o)=1$) and the even signature; these characteristics of the Pomeron are, of course, assumed for precisely that reason.

The simple Regge parameterization thus yields two characteristic results: Constant total cross sections, and factorization of total cross-sections. Neither of these features contradicts experiment; conversely, experiment does not, as yet, provide unequivocal support of them either.

The prediction of factorization, in fact, raises some theoretical problems.[2)] For scattering of one nucleus of mass number A on another of mass number A', simple intuition suggests a cross-section varying with A and A' like $(A^{1/3}+A'^{1/3})^{2/3}$. This just says that the cross-section for two black objects is proportional to the area swept out. Evidently, such a behavior is incompatible with factorization. Regge behavior in general corresponds to scattering from a structure whose radius grows and which becomes increasingly transparent; such a picture means a cross section varying like AA'. Within the multiperipheral model this growth of the radius is connected to the growing number of particles produced, as will become evident in chapter 16. One may then no longer think of the nucleus, or, for that matter, of the nucleon, as an object with a fixed size.

If the Pomeron really is just a Regge pole, the AA' behavior is inescapable. If it is in actuality some sort of j-plane branch point, or combination of branch points, then anything is possible. A branch point singularity may imply factorization (that follows automatically from unitarity for any kind of j-plane singularity if the singularity occurs in only a single eigenchannel) but it need not imply factorization.

The prediction that cross sections are constant provides us with an immediate test of the nature of the Pomeron singularity. Since cross sections are limited by the Froissart bound (appendix B), a power behavior, which is implied by the Pomeron pole picture, necessarily means that the upper bound is a constant. Any ℓn s deviation from this constant corresponds to a more complicated j-plane structure. In appendix D we give a detailed summary of the j-plane language. As can be seen there a $\ell n\, s$ or $\ell n^2 s$ increase of σ_T is caused by a double or triple pole respectively at $\alpha(o)=1$. Such higher order poles can also be some limiting case as $t\to 0$ of a complicated cut structure. The measured total cross sections (see chapter 1) tell us that such a structure may very well exist but its present effects are relatively small.

In addition to the Pomeron we encounter lower lying Regge trajectories with $\alpha_R(t)\approx \frac{1}{2}+t$. Their effect on the total cross section is to introduce an energy dependent correction of the type

$$\sigma_{AB} = \beta_A\beta_B + \sum_i \beta_{A_i}\beta_{B_i}\tau_i s^{\alpha_i(o)-1} \tag{5.5}$$

where we assume that $\alpha_P(o)=1$ and introduce a sum over all lower trajectories. τ_i are the signatures of these trajectories. The impact of such corrections can be seen in fig. 1.1 in the low s region. In particular we note that their coupling in processes that have exotic s-channel quantum numbers is negligible, and that in all other measured cases their coupling is positive. This fact is a cornerstone of the "two-component duality" picture[3] that associates the secondary Regge trajectories with s-channel resonance formation, and relates the t-channel Pomeron to s-channel background scattering. In this case, since each resonance contributes positively to σ_T, it is obvious that the coupling of the energy dependent term is positive and σ_T approaches its asymptotic behavior from above.

Although we do not intend to digress into a lengthy discussion of these lower lying Regge trajectories we would like to mention one important fact that stems from the Serpukhov experiments - namely that they seem to be much closer to pure poles than does the Pomeron. This is evident from the fact that differences of total cross sections within the same isospin multiplet continue to decrease with a power of s throughout the Serpukhov energy range in much the same way that they behaved at lower energies. These data are shown in fig. 5.1. This smooth continuation should be contrasted with the obvious change in behavior of the total cross sections themselves as shown in fig. 1.1, which suggests that the Pomeron is different from a pure Regge pole.

The Pomeron pole picture has characteristic implications for elastic differential cross sections as well as for total cross sections, as is evident from eq. (5.1).

First of all, we note that our assumptions of even signature and intercept unity produce a purely imaginary elastic amplitude at t=0. The ratio of real to imaginary parts of the elastic amplitude, near t=0, can be expressed in terms of the slope α' of the Pomeron: We write

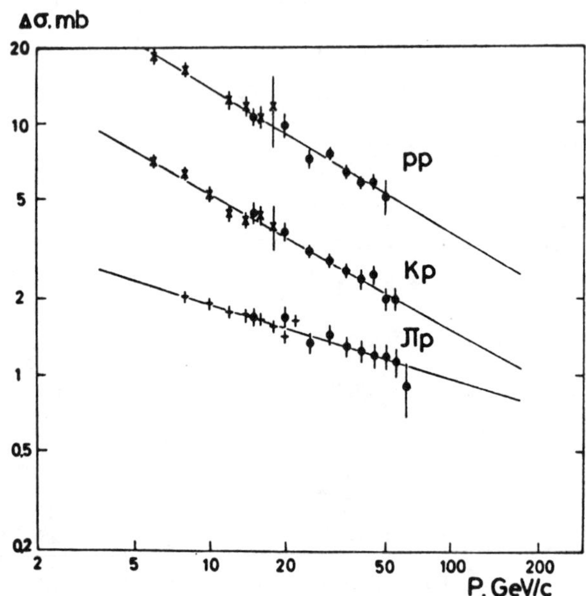

Fig. 5.1. Differences between particle and anti-particle total cross sections. The data are the same as those appearing in fig. 1.1. Taken from ref. 4. The shown fits are of the form p^{-n} with $n=0.61\pm0.03$, 0.56 ± 0.02, 0.31 ± 0.04 for pp, Kp and πp respectively.

$$\alpha(t) = 1 + \alpha' t$$

and hence from (5.1) we see that

$$\frac{\mathrm{Re}T(s,t)}{A(s,t)} = \frac{\pi}{2}\,\alpha' t + \ldots \tag{5.6}$$

for small t. Thus, the flatter the Pomeron, the less real part there is.

The differential cross section for elastic scattering is, from (5.1),

$$\frac{d\sigma}{dt} = \frac{1}{16\pi s^2} \, |\beta_A(t)\beta_B(t)(s)^{\alpha(t)} \cdot$$

$$\cdot \left(- \frac{1 + e^{-i\pi\alpha(t)}}{\sin\pi\alpha(t)}\right)|^2 \tag{5.7}$$

so that

$$(d\sigma/dt)_{t=0} = \frac{\beta_A{}^2 \beta_B{}^2}{16\pi} \tag{5.8}$$

and

$$\frac{(d\sigma/dt)}{(d\sigma/dt)_{t=0}} = e^{2\alpha' t \ \ell n(s)} \left|\frac{\beta_A(t)\beta_B(t)}{\beta_A\beta_B}\right|^2 \tag{5.9}$$

for small t. We thus have a slope parameter b(s) containing a term proportional to ℓn s. If we guess the residue functions to be exponential for small t, and write

$$\beta_A(t) = \beta_A \ e^{b_A t}$$

and

$$\beta_B(t) = \beta_B \ e^{b_B t}$$

then the slope parameter is

$$b(s) = 2b_A + 2b_B + 2\alpha' \ \ell n \ s. \tag{5.10}$$

We have commented before on the experimental situation with regard to b(s): It seems clear that α' for the Pomeron is no more than 0.3 or so,[5)] which makes the Pomeron (if it is just a Regge pole) rather different from other Regge trajectories.

The possibility that α' is exactly zero, and that $\alpha(t)\equiv 1$ - a flat Pomeron - is not incompatible with the data and deserves special mention. We may take the limit of (5.1) as $\alpha(t)\to 1$ to see that in this event

$$T_{AB\to AB}(s,t) = is\beta_A(t)\beta_B(t) \tag{5.11}$$

and that therefore

$$\frac{d\sigma}{dt} = \frac{1}{16\pi}|\beta_A(t)\beta_B(t)|^2. \tag{5.12}$$

We must keep in mind, however, that a flat trajectory $\alpha(t)=1$ corresponds to a fixed pole at $j=1$ in the crossed channel partial wave amplitude, and as is explained in appendix D, this situation cannot be maintained up to values of t above t-channel threshold ($t=4m_\pi^2=0.08$ GeV^2 in practice) without violating t-channel unitarity.[6] Something must occur to prevent the continuation; one possibility is a j-plane branch point. If this is a square root branch point and passes through $j=1$ at $t=4m_\pi^2$, then a fixed pole Pomeron can be shunted onto an unphysical sheet of the j-plane for $t>4m_\pi^2$, and unitarity can be saved.[7,8] An example of an amplitude with these properties is given in appendix D.

A flat Pomeron with $\alpha(t)=1$ also gives us trouble in inclusive processes, as we shall see in the following two chapters. We shall also point out in chapter 8 that it will cause trouble in multibody exclusive processes as well. Again, it is possible that j-plane branch points can save the situation; but at the very least these will cause our simple Regge picture to break down.

Other versions of nearly flat Pomerons may be invented. One example is a pair of trajectories, complex conjugates for $t<0$, with real part equal to one:[9)] For example, take

$$\alpha_{\pm}(t) = 1 \pm \alpha' \sqrt{t\ (4m_{\pi}^{2}-t)}. \qquad (5.13)$$

For $t>4m_{\pi}^{2}$ and $t<0$, $\alpha_{\pm}$ are complex; hence there is no trouble with t-channel unitarity. Nevertheless, $\mathrm{Re}\,\alpha_{\pm}(t) = 1$ for $t<0$. The existence of a branch point in $\alpha(t)$ at $t=0$, as postulated in (5.13), does not imply such a branch point in $T(s, t)$ itself, since the trajectory occurs as a complex conjugate pair. The dynamical origin of a branch point in $\alpha(t)$ is obscure, but could be connected to the existence of a j-plane cut passing through $j=1$ at $t=0$ with which the Pomeron interacts, or "collides."[10)]

The differential cross section resulting from a Pomeron pair like this is

$$\frac{d\sigma}{dt} = \frac{1}{16\pi}\left|\beta_A(t)\beta_B(t)\right|^2 \Big\{1 + \sinh^2\!\left(\frac{\pi}{2}\,\mathrm{Im}\,\alpha(t)\right) - \sin^2\!\left(\mathrm{Im}\,\alpha(t)\,\ln s\right)\Big\}; \qquad (5.14)$$

it contains dips at $\mathrm{Im}\alpha(t)\ell n\ s=n\pi$, $n=0, 1, 2 \ldots$ Evidently, these dips move in to smaller values of $|t|$ as s increases. The diffraction peak is, therefore, not really nonshrinking.

So much for flat, or nearly flat, Pomerons. Modifications of the simple Regge formula for moving Pomerons are also possible, and are frequently invoked. The most common modification is to introduce j-plane branch points. There are some explicit dynamical models giving rise to such corrections, as we shall see in a later chapter. At this point we content ourselves with pointing out some of the phenomenological features that cuts might exhibit.

In general, an even signature j-plane cut may be expected to contribute to T(s, t) a term of the form (see appendix D)

$$T_c(s,t) = \gamma_c(t)(-e^{-\frac{i\pi}{2}\alpha_c(t)})(s)^{\alpha_c(t)} \cdot$$

$$\cdot \left(\ell n \frac{s}{s_o} - \frac{i\pi}{2}\right)^{-\nu} \tag{5.15}$$

to be added to the Pomeron pole (5.1). If $\alpha_c(0)=1$, this adds to the total cross section a term behaving like $(\ell n\ s)^{-\nu}$. If $\alpha_c(t)>\alpha(t)$ (for $t<0$), the cut will eventually dominate the Pomeron pole, in spite of the $(\ell n\ s)^{-\nu}$ factor. The dominant s-behavior will then be associated with $\alpha_c(t)$ rather than with $\alpha(t)$. If the cut and pole are comparable, interference effects will occur; destructive interference (γ of opposite sign to $\beta_A \beta_B$) will lead to dips in $d\sigma/dt$. Such dips are, as we have seen, observed and are sometimes attributed to the presence of cuts.

It is evident that if cuts are indeed important, the predictive power of the simple Regge mnemonic almost disappears. Unless the cuts and at least some of their properties can be deduced theoretically, from the Regge poles or from elsewhere, one may as well abandon attempts to parameterize data with Reggeism.

6
One-Particle Inclusive Distributions: Mueller Analysis

The next simplest cross sections to discuss are one-particle inclusive cross sections. We are concerned with the reaction $A+B \to C+\text{anything}$, for which the cross section is $d\sigma^C_{AB}/d^3\vec{q}_C$. We define the one-particle distribution by

$$\rho^C_{AB} = \omega_C \frac{d\sigma^C_{AB}}{d^3 q_C} . \tag{6.1}$$

The distribution ρ depends on three independent variables, which we shall commonly choose from among

$$\begin{aligned} s &= (p_1 + p_2)^2 \equiv P^2 \\ t &= (p_1 - q)^2 \\ u &= (p_2 - q)^2 \\ M^2 &= (p_1 + p_2 - q)^2 . \end{aligned}$$

Evidently, these are not independent; we have $s+t+u=M^2+m_A{}^2+m_B{}^2+m_C{}^2$. These variables are the natural ones in terms of which to introduce a Regge parameterization. The variables are displayed in fig. 6.1.

We recall the generalized optical theorem (appendix A), which states that

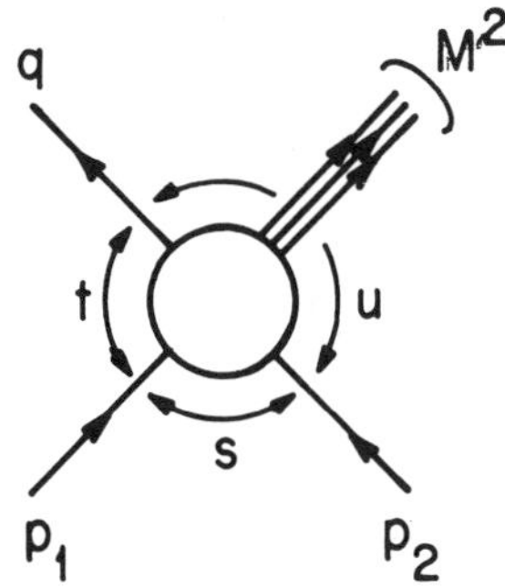

Fig. 6.1

$$\rho^{C}_{AB} = \frac{1}{s} A_{A\bar{C}B \to A\bar{C}B}\Big|_{\text{forward}}, \tag{6.2}$$

and which we illustrate schematically in fig. 6.2.

To apply eq. (6.2), and the Regge picture, to one-particle inclusive distributions we must distinguish the three separate kinematic regions defined in chapter 2, namely, the projectile and target fragmentation regions and the central region. In each of these regions the variables s, t, u, and M^2 look different.

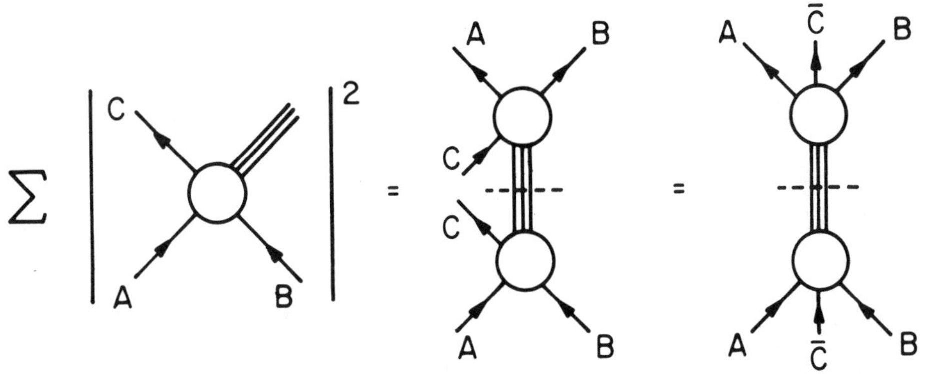

Fig. 6.2

(i) Projectile fragmentation region

This region is characterized by $Y/2 - y$ finite, where Y is the overall rapidity range and y the c.m. rapidity of the produced particle. We have approximately,

$$Y = \ln \frac{s}{m_T^2}$$

and from eq. (2.21)

$$y \simeq \ln \frac{x\sqrt{s}}{m_T} \qquad (y>>1)$$

and therefore

$$\frac{Y}{2} - y \approx - \ln x; \tag{6.3}$$

hence finite $\frac{Y}{2} - y$ corresponds to $x>0$.

Now we have

$$t = (p_1 - q)^2$$

$$\approx m_A^2+m_C^2- \frac{2m_A^2}{\sqrt{s}} m_T \sinh y - m_T\sqrt{s}\, e^{-y} \tag{6.4}$$

and $$u \approx m_B^2+m_C^2- \frac{2m_B^2}{\sqrt{s}} m_T \sinh y - m_T\sqrt{s}\, e^{y}. \tag{6.5}$$

Thus if $Y/2-y$ is finite, we conclude that t is finite and u is of order s. Projectile fragmentation corresponds to holding the momentum transfer from A to C fixed while that from B to C grows with energy.

We also have

$$M^2 = (P - q)^2$$

$$= s + m_C{}^2 - 2\sqrt{s}\, m_T \cosh y. \qquad (6.6)$$

Hence if $y \sim Y/2$ we have M^2 of order s. As already mentioned in eq. (2.20)

$$M^2 = s(1 - x) + 2t. \qquad (6.7)$$

This means that over most of the region, $M^2 \sim s$; however, the region also includes a subregion where $M^2/s \to 0$, because then we can arrange the two terms in eq. (6.6) to cancel. It corresponds, therefore, to the extreme edge of the rapidity range, where the produced particle is going nearly as fast as the projectile.

(ii) Central region

Here we have y finite - (i.e., both $Y/2-y$ and $Y/2+y$ grow with Y). Hence, from eqs. (6.4) and (6.5), we find

$$t \approx -\sqrt{s}\, m_T\, e^{y}$$
$$u \approx -\sqrt{s}\, m_T\, e^{-y} \qquad (6.8)$$

so that both t and u grow, with

$$tu \approx m_T{}^2 s \quad \text{and} \quad M^2 \to s. \qquad (6.9)$$

The condition under which (6.9) holds can also be expressed in terms of x; it is $x^2 << m_T{}^2/m_A m_B$.

(iii) Target fragmentation region

We now have $Y/2+y$ finite; the behavior is the same as in the projectile fragmentation region but with t and u interchanged. We have $t \sim s$, u fixed, and

$$M^2 = s(1 + x) + 2u. \qquad (6.10)$$

Thus generally $M^2 \sim s$, but there is also a subregion where $y \to -Y/2$, or $x \to -1$, in which $M^2/s \to 0$.

Now our general rule for invoking Regge behavior is that if anything gets large, then in the channel crossed to that variable there is a leading Regge pole, and the asymptotic behavior is determined by that trajectory.

Let us first apply this principle to $A_{A\bar{C}B \to A\bar{C}B}$ in the projectile fragmentation region. (The target fragmentation region is, obviously, similar.) Here s and $u \to \infty$ while t remains fixed. Thus, as illustrated in fig. 6.3, we should expect

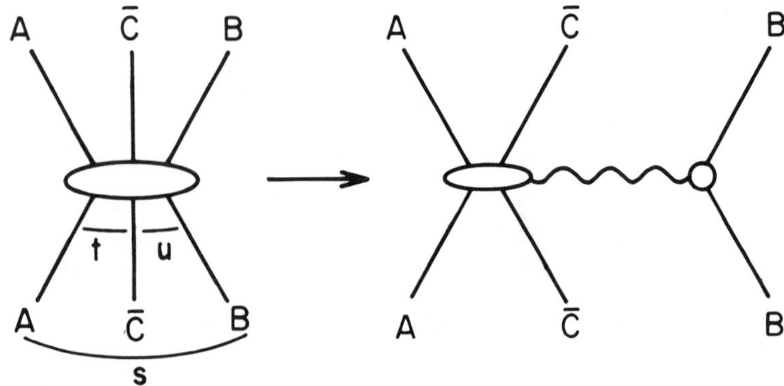

Fig. 6.3

$$A_{A\bar{C}B\to A\bar{C}B}\Big|_{\text{forward}} \to \beta_{AC}(q_T, \tfrac{Y}{2}-y)\nu^{\alpha(o)}\ \beta_B \qquad (6.11)$$

The variable ν can be either $s=(p_1+p_2)^2$ or M^2 or $-u=-(-q+p_2)^2$. Since all are of the same order it is just a matter of convenience which we choose. We will take $\nu=s$. The trajectory α is associated with a Pomeron pole which is assumed to factorize as implied by this formula. Here β_B is just the usual Pomeron coupling to the physical particle B; this coupling is evaluated at zero momentum transfer because we want the forward 3→3 amplitude. The "vertex" β_{AC} describes the coupling of the Pomeron to the structure containing A and C shown in the left of fig. 6.3. This depends, clearly, on q_T and $Y/2-y$.

In order to make this formula more plausible to the reader who is not acquainted with this approach let us sketch how it arises in the multiperipheral model (to be discussed in detail in part IV). The Pomeron pole in the total cross-section is described by the diagrams in fig. 6.4 where the broken line designates the discontinuity

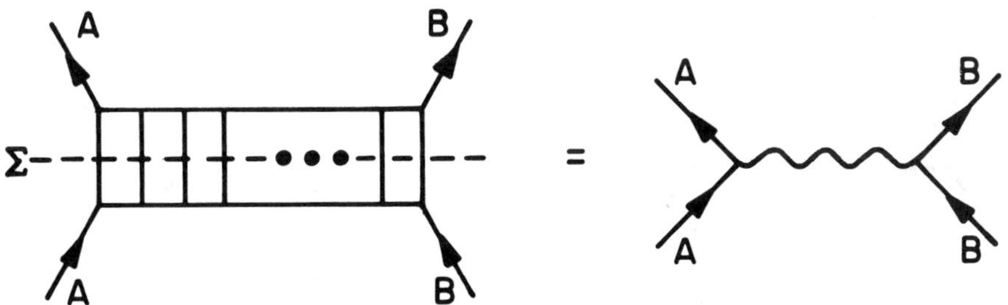

Fig. 6.4

in s. The sum of ladder diagrams builds up a factorizing Pomeron. In studying an inclusive process one breaks one of the rungs of this ladder which, for the fragmentation problem at hand, leads to the diagrams in fig. 6.5. This amounts to the statement implied above with the Pomeron being once again represented by a sum over all ladder diagrams. This leads to the equation

$$\rho^C_{AB} = \beta_{AC}\ (q_T, \frac{Y}{2}-y)\ \beta_B\ s^{\alpha(o)-1} . \qquad (6.12)$$

The factorization of the Pomeron means that β_B is the same term that appears in the equation

$$\sigma_{AB} = \beta_A \beta_B\ s^{\alpha(o)-1} . \qquad (6.13)$$

We learn, therefore, two things. First, the equality $\alpha(o)=1$ implies in eq. (6.12) that scaling is obtained. Secondly, factorization of the Pomeron pole leads to

$$\frac{1}{\sigma_{AB}}\ \rho^C_{AB} = \frac{\beta_{AC}(q_T, \frac{Y}{2} - y)}{\beta_A} \qquad (6.14)$$

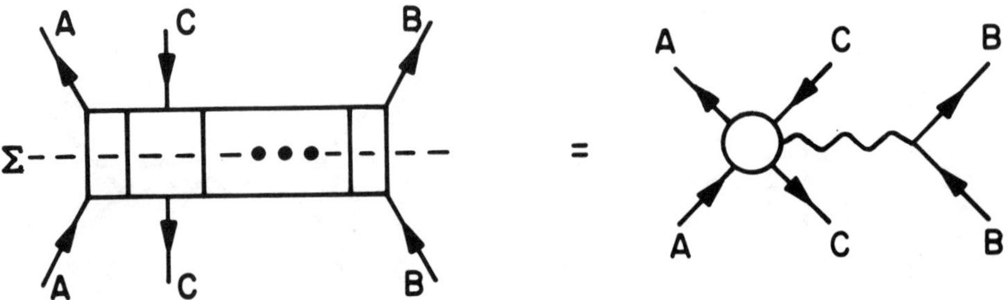

Fig. 6.5

a quantity which is independent of particle B. Thus the simplest Regge pole representation establishes the fragmentation picture of rules 3 and 4 of chapter 2.

What should we expect from the central region? Here t and u are both large, so as illustrated in fig. 6.6 we should have

$$A_{A\bar{C}B\to A\bar{C}B}\Big|_{\text{forward}} \to \beta_A(-t)^{\alpha(o)}\beta_C(-u)^{\alpha(o)}\beta_B. \tag{6.15}$$

We recall that $tu=(q_T{}^2+m_C{}^2)s$ in this region; thus we find

$$\rho_{AB}^{C} \to \beta_C(q_T)s^{\alpha(o)-1}\beta_A\beta_B \text{ and } \frac{1}{\sigma_{AB}}\rho_{AB}^{C} \to \beta_C(q_T). \tag{6.16}$$

There is now no dependence on y; β_C can be a function only of q_T. As in the fragmentation regions, since $\alpha(o)=1$, there is no s-dependence. Furthermore, factorization suggests that $\frac{1}{\sigma_{AB}}\rho_{AB}^{C}$ is independent of both A and B. We will return to this

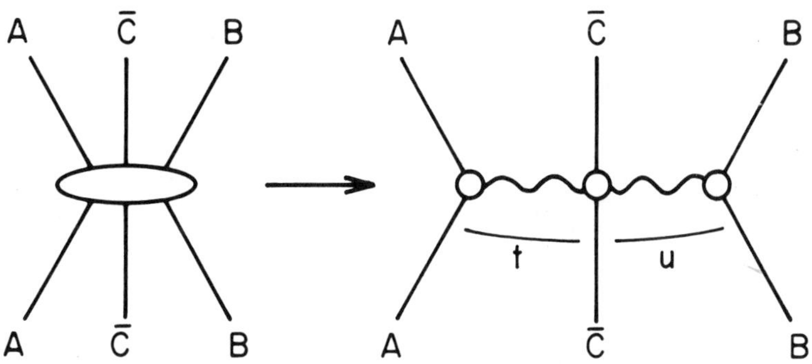

Fig. 6.6

point in fig. 6.8 below.

Thus Reggeism would imply that there is a central plateau which is unaware of either end of the rapidity distribution, but depends only on the type of produced particle and its q_T. Statements about the properties of the central region can now be made in terms of a 2-Pomerons - 2-particles coupling. Thus rule 7 of chapter 2 means that the 2-Pomerons - 2-pions coupling is much larger than any other coupling of this kind. SU_3 breaking does not imply that the Pomeron is not an SU_3 singlet - it can be understood as a symmetry breaking of the vertex 2-Pomerons - 2-particles.

Evidently, the Regge mnemonic reproduces for us all of the important experimental properties which we identified earlier, without invoking anything except the Pomeron trajectory.[11] Scaling and

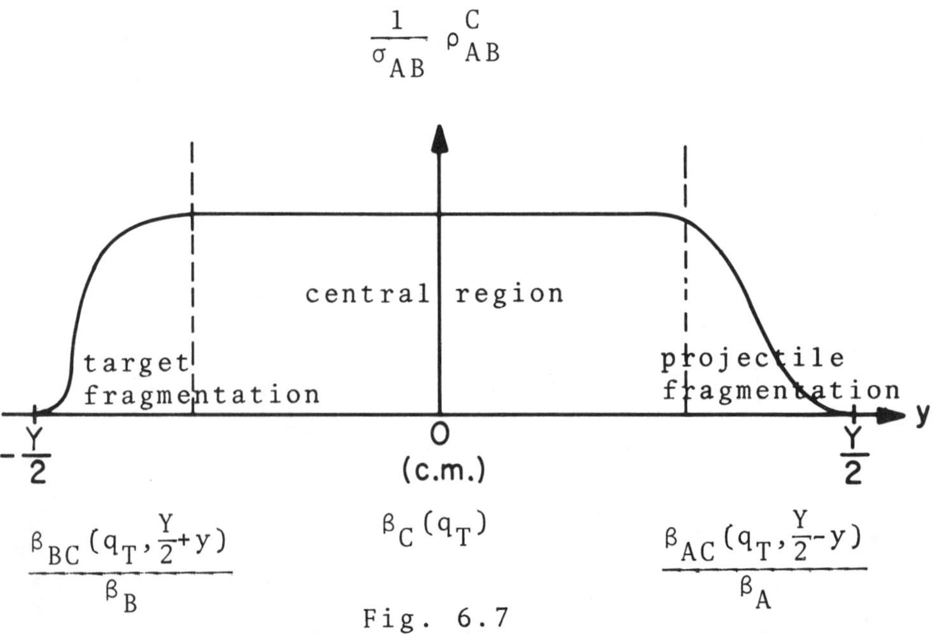

Fig. 6.7

limiting fragmentation are consequences only of intercept unity and factorization. The rapidity plot to be expected is schematically summarized in fig. 6.7, with the shape determined by the indicated Pomeron vertex functions.

At less than asymptotic energies, lower lying Regge trajectories can be expected to contribute in the various regions, in addition to the leading Pomeron. Thus, for example, in the projectile fragmentation region, we should expect a correction to eq. (6.12) of the form

$$\rho^{C}_{AB} \to \beta_B \beta_{A\bar{C}}(q_T, \tfrac{Y}{2}-y) + \sum_i \beta^i_{A\bar{C}}(q_T, \tfrac{Y}{2}-y)\, \beta^i_B(o) \cdot$$

$$\cdot \tau_i s^{\alpha_i(o)-1} \qquad (6.17)$$

coming from lower lying trajectories $\alpha_i(t)$, which couple to the particle B with β^i_B and to the structure $A\bar{C}$ with $\beta^i_{A\bar{C}}(q_T, \frac{Y}{2}-y)$.

In terms of rapidity, such an extra term is proportional to exp $\left[Y(\alpha_i(o)-1)\right]$ which can be rewritten as $\exp(-Y/L)$ where $L=(1-\alpha_i(o))^{-1}$.[12)] L is called the correlation length and will play a key role in the discussion of two particle inclusive distributions in the next chapter. Since lower-lying trajectories have intercepts which are characteristically about 1/2, we should have $L \approx 2$. This conclusion is also reasonably well in accord with experiment.

In two-body processes, duality suggests to us that exotic reactions, which have no resonances, also have no exchanged Regge trajectories other than the Pomeron.[3] They, therefore, are expected to approach their asymptotic value more quickly than nonexotic reactions.

The projectile fragmentation region corresponds to the scattering of the structure $A\bar{C}$ from the particle B. The "two-body process" of interest is this scattering; namely $(A\bar{C})+B\to(A\bar{C})+B$. Thus one might expect, through duality, a rapid approach to scaling if the channel $A\bar{C}B$ is exotic.*) Moreover, we may expect the additional terms in (6.17) to be positive if their duality to resonances holds. In fact, one does indeed find that the inclusive invariant amplitudes approach the scaling limit from above in the fragmentation region. A particular example of this behavior is shown by the $(\pi^+p, \pi^\pm)$ data in fig. 2.3. The $(\pi^+ p, \pi^-)$, an exotic channel, shows early scaling, and (π^+p, π^+) exhibits a decrease towards the scaling limit. Until NAL data become available we have to use these lower energy data to test these ideas.

An analysis in terms of conventional Regge poles can lead to a useful parameterization of the

*Since several variables play a role here, this condition may not be sufficient. Thus one may have to require that also AB be exotic. This seems natural in view of the fact that the integral over ρ_{AB}^{C} is related to σ_{AB}. More restrictions may be required and we refer to table II in ref. 13 for a summary of recent discussions of this point.

inclusive data, and the approach to scaling.[14] Thus, in ref. 14, the authors considered the effects of exchange degenerate f, ρ, ω, and A_2 trajectories. The usual exchange degeneracy conditions

$$\beta^f_{\pi^+} = \beta^\rho_{\pi^+}, \qquad \beta^f_p = \beta^\omega_p, \qquad \beta^\rho_p = \beta^{A_2}_p,$$

$$\beta^f_{K^+} = \beta^\omega_{K^+} = \beta^\rho_{K^+} = \beta^{A_2}_{K^+} \tag{6.18}$$

can be used in explaining the behavior of the experimental distributions. The early scaling of (pp, π^-) (K^+p, π^-) and (π^+p, π^-) in the target frame, as observed in fig. 2.4, leads to

$$\beta^f_{p\pi^-} = \beta^\omega_{p\pi^-} = \beta^\rho_{p\pi^-} = \beta^{A_2}_{p\pi^-} \tag{6.19}$$

thus exhibiting an exchange degenerate pattern. However, the (π^-p, π^-) contains a correction to scaling since $\beta^f_{\pi^-} = -\beta^\rho_{\pi^-} = \beta^f_{\pi^+}$. Similar relations can be found for other vertices. Thus from either experiment or theoretical reasoning one can find

$$\beta^\rho_{\pi^+\pi^-} \approx \beta^f_{\pi^+\pi^-}, \qquad \beta^f_{p\pi^+} \approx \beta^\omega_{p\pi^+}, \qquad \beta^\rho_{p\pi^+} \approx \beta^{A_2}_{p\pi^+},$$

$$\beta^\rho_{K^+\pi^-} \approx \beta^\omega_{K^+\pi^-} \approx \beta^f_{K^+\pi^-} \approx \beta^{A_2}_{K^+\pi^-} \tag{6.20}$$

which establish a pattern reminiscent of the vertex functions of (6.18).

Corrections due to lower lying Regge trajectories also show up in the central region. Here we would expect eq. (3.12) to be modified to read

$$\rho_{AB}^{C} \to \beta_A \beta_B \beta_C (q_T)$$

$$+ \sum_i \beta_C^{iP} (q_T) \left(\beta_B \beta_A^i \frac{(-t)^{\alpha_i(o)} (-u)}{s} \right.$$

$$\left. + \beta_A \beta_B^i \frac{(-t)(-u)^{\alpha_i(o)}}{s} \right) +$$

$$+ \sum_{i,j} \beta_C^{ij} (q_T) \beta_A^i \beta_B^j \frac{(-t)^{\alpha_i(o)} (-u)^{\alpha_j(o)}}{s} \ . \qquad (6.21)$$

The successive terms here correspond to double Pomeron exchange, to one Pomeron and one lower-lying trajectory exchange, and to double exchange of the lower trajectories.

Since $t \sim u \sim \sqrt{s}$ in the central region, these terms behave, successively, like 1, $\sqrt{s}^{\alpha_i(o)-1}$ and $s^{\alpha_i(o)-1}$; that is, if $\alpha_i(o) \sim 1/2$, like 1, $s^{-1/4}$ and $s^{-1/2}$. Thus in this region, the last correction to die away disappears like $s^{-1/4}$, in contrast to the $s^{-1/2}$ behavior of the correction in the fragmentation regions. Since $t \sim e^{-(Y/2-y)}$ and $u \sim e^{-(Y/2+y)}$, in terms of rapidity the corrections take the form

$$\rho_{AB}^{C} \to \beta_A \beta_B \beta_C + \sum_i \beta_A^i \beta_B \beta_C^{iP} e^{-(Y/2-y)/L_i}$$

$$+ \sum_i \beta_A \beta_B^i \beta_C^{iP} e^{-(Y/2+y)/L_i}$$

$$+ \sum_{i,j} \beta_A^i \beta_C^{ij} \beta_B^j \ e^{-Y/L_{ij}} \qquad (6.22)$$

where, as before, $L_i = 1/(1-\alpha_i(o))$, and where $L_{ij} = 1/(1-\alpha_i(o)-\alpha_j(o))$.

Eq. (6.21) also relates the correction to scaling (fixed x, varying s) to the approach to the central region from the fragmentation regions (fixed s, varying x). Thus the term $\frac{(-t)^{\alpha_i(o)}(-u)}{s}$ leads, at x=0, to $s^{-1/4} m_T^{3/2}$ whereas for fixed s we can rewrite it in the slightly positive x region as $m_T\, x^{\alpha_i(o)}$.

Experimentally, it turns out that the inclusive distributions in the central region increase with energy[15)] and one may even go so far as to say that they are consistent with an $s^{-1/4}$ behavior and aim at the same asymptotic limit. This view is substantiated by fig. 6.8. Alternatively, it may be that the observed increase is some thresh-

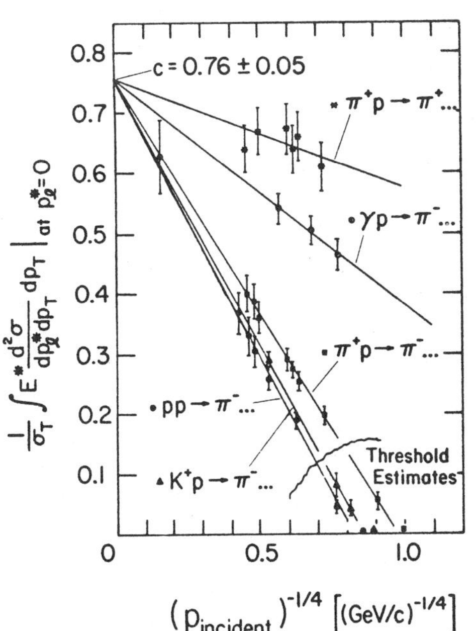

Fig. 6.8. Approach to scaling in central region. Figure taken from ref. 15. For normalization by σ_T the following values were used: $\sigma_T(pp) = 39.8$ mb, $\sigma_T(\pi^+p) = 23.4$ mb, $\sigma_T(K^+p) = 17.4$ mb, $\sigma_T(\gamma p) = 99$ μb.

old effect and an $s^{-1/4}$ decrease will be encountered later on. If the increase persists it breaks, for the first time, the analogy with naive two-component duality.[16)] The latter would predict that the Regge corrections are due to resonances and are, therefore, positive thus leading to a decrease both of $\rho^{C}_{AB}(s, x)$ when $x \to 0$, as well as a decrease of $\rho^{C}_{AB}(x=0)$ when s increases.

At this point it may be worthwhile to point out again the extreme assumptions implied by using the Regge pole language. One can easily conceive of many types of models which will not lend themsleves to such a simple analysis as implied by eq. (6.21). Let us look at some examples as shown in fig. 6.9. The first, (fig. 6.9a), is just a correction to the leading multiperipheral contribution which is usually alleged to be negligible - a statement which is true only asymptotically. The second example, (fig. 6.9b), is a nonplanar graph - a polyperipheral diagram. This is the kind that

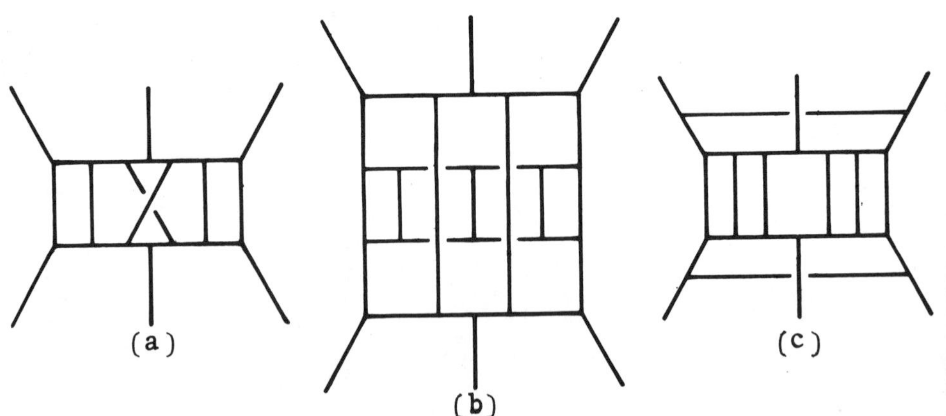

Fig. 6.9

leads to Mandelstam cuts (see the discussion in chapter 10) and is supposed to lag behind the pole by one power of ℓn s. Finally, we look in fig. 6.9c at a characteristic absorptive model. All these are counter-examples to the simple representation of fig. 6.6; nevertheless, they will lead to a plateau in rapidity with possible overall ℓn s factors. Hence the apparent existence of the plateau does not verify the correctness of fig. 6.6, and does not imply that the other conclusions like eqs. (6.16) and (6.17) are true. Therefore, in applying these equations to experimental analyses fair amounts of optimism are needed.

We mentioned that the edges of the rapidity distributions, where $y \to \pm\frac{Y}{2}$, are characterized equally well by the limits $x \to \pm 1$ or by $M^2/s \to 0$ with either t or u fixed. The relevant kinematic region is, therefore, $s \gg M^2 \gg m^2$. These same regions can also be attained by a different sequence of limits; namely, first let $s \to \infty$ with M^2 and t or u fixed and then let $M^2 \to \infty$.

If the two limiting procedures are equivalent, then we may use a different version of the simple Regge rules to find out how the functions $\beta_{AC}(q_T, \frac{Y}{2}-y)$ and $\beta_{BC}(q_T, \frac{Y}{2}+y)$ behave as $y \to \pm\frac{Y}{2}$.[17)]

Let us take s to infinity with M^2 and t fixed. Then we may think of the amplitude for the inclusive reaction $A+B \to C+$anything as the "two body" process $A+B \to C+X$ where the "particle" X has mass M.

The cosine of the center of mass scattering angle for the crossed reaction, $A+\overline{C}\to\overline{B}+X$, is proportional to s/M^2; hence the usual Regge rules tell us that

$$\frac{d\sigma^C_{AB}}{dt\; dM^2} \xrightarrow[\substack{\frac{s}{M^2}\to\infty \\ t, M^2 \text{ fixed}}]{} \frac{1}{16\pi s^2}\left|\sum_i \beta^i_{AC}(t)\left(\frac{s}{M^2}\right)^{\alpha_i(t)} \cdot \beta^i_B(t,\, M^2)\,\frac{-1+\tau_i e^{-i\pi\alpha_i}}{\sin\pi\alpha_i}\right|^2 . \qquad (6.23)$$

Here $\alpha_i(t)$ is a Regge trajectory in the $A\overline{C}$ channel; $\beta^i_{AC}(t)$ is its coupling to A and C; and $\beta^i_B(t,\, M^2)$ represents its coupling to B and the "particle" X, with the dependence on the mass of X explicitly displayed.

The limit is displayed graphically in fig. 6.10.

We next wish to take the limit $M^2\to\infty$. To do this we need the behavior of $|\beta^i_B(t,\, M^2)|^2$ for large

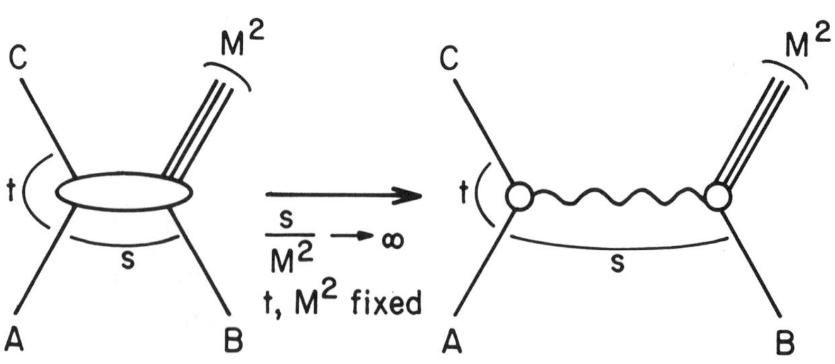

Fig. 6.10

M^2. What we can expect to happen is illustrated in fig. 6.11; $|\beta_B(t, M^2)|^2$ is the "total cross section" for the Reggeon $\alpha_i(t)$ to scatter from the particle B at total c.m. energy M^2. As $M^2 \to \infty$, it should, therefore, be dominated by Pomeron exchange. Thus we presume

$$|\beta_B^i(t, M^2)|^2 \to \beta_{RRP}^i(t)(M^2)^{\alpha(o)}\beta_B. \qquad (6.24)$$

In this equation α is the Pomeron, β_B is its coupling to B and $\beta_{RRP}(t)$ is its coupling to the Reggeon $\alpha(t)$; this last factor is known as the triple Regge vertex for Pomeron-Reggeon-Reggeon. We have been cavalier in treating the signature here; we shall remedy this shortly.

Finally, we recall the kinematic relation

$$\omega \frac{d\sigma}{d^3q} = \frac{s}{\pi} \frac{d\sigma}{dt\,dM^2}.$$

Putting all this together we see that

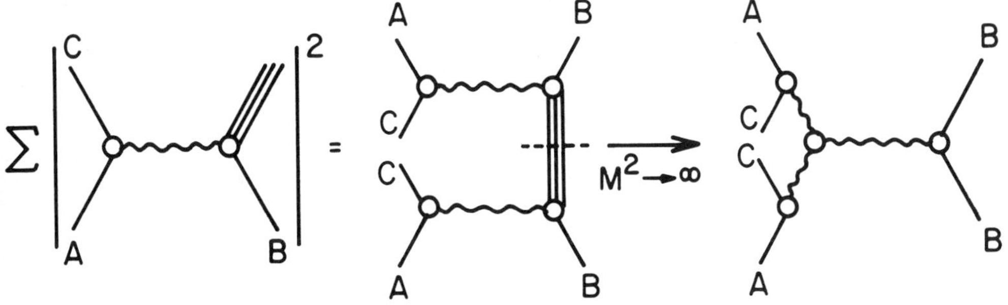

Fig. 6.11

$$\rho_{AB}^{C} \to \frac{1}{16\pi^2} \frac{|\beta_{AC}(t)|^2}{\beta_A} \left(\frac{s}{M^2}\right)^{2\alpha_R(t)-1} \cdot$$

$$\cdot \; (M^2)^{\alpha(o)-1} \beta_{RRP}(t), \tag{6.25}$$

as $\frac{s}{M^2} \to \infty$ first and then $M^2 \to \infty$.

Now we wish to assume this to coincide with what we obtain from the limit first s, $M^2 \to \infty$ and then $\frac{M^2}{s} \to 0$. That gave

$$\rho_{AB}^{C} \to \frac{\beta_{AC}(q_T, \frac{Y}{2}-y)}{\beta_A} \; s^{\alpha(o)-1} \tag{6.26}$$

with $y \to \frac{Y}{2}$; (i.e., with $x \to 1$.) We first note that $\alpha(o)=1$; we next note that $M^2/s = 1-x$. Thus the two expressions are compatible provided that

$$\beta_{AC}(q_T, \frac{Y}{2}-y) \sim (1-x)^{1-2\alpha_R(t)} \tag{6.27}$$

as $x \to 1$.[17]

Everything proceeds the same way if we hold u instead of t fixed. Again the two expressions are compatible provided

$$\beta_{BC}(q_T, \frac{Y}{2}+y) \sim (1+x)^{1-2\alpha_R(u)} \tag{6.28}$$

as $x \to -1$, where $\alpha_R(u)$ now stands for the leading trajectory in the $B\overline{C}$ channel.

Reggeism thus specifies the 1-x, or $\frac{Y}{2}-y$, dependence of β_{AC}. The q_T dependence of β_{AC} as

$x \to 1$ is related to the t-dependence in $|\beta_{AC}(t)|^2$ and in $\beta_{RRP}(t)$. The function $\beta_{AC}(t)$ is just a conventional Regge residue function describing the coupling of $\alpha_R(t)$ to A and C. The triple Regge vertex is, on the other hand, a new entity which has not been encountered in conventional two-body Regge phenomenology.

Evidently, triple Reggeon vertices can be defined for the coupling of any three Regge trajectories, and in principle such vertices can be studied experimentally by subtracting off the effects of leading Regge trajectories in the triple Regge region to expose lower trajectories. Thus, including lower lying stuff, we expect (6.25) to be replaced by the statement

$$\rho^C_{AB} \xrightarrow[\substack{s/M^2\to\infty \\ M^2\to\infty}]{} \frac{1}{16\pi^2} \sum_{i,j,k} \beta^i_{AC}(t)\beta^{j*}_{AC}(t) \left(\frac{s}{M^2}\right)^{2\alpha_i(t)-1} \cdot \left(-\frac{1+\tau_i e^{-i\pi\alpha_i(t)}}{\sin\pi\alpha_i(t)}\right)\left(-\frac{1+\tau_j e^{-i\pi\alpha_j(t)}}{\sin\pi\alpha_j(t)}\right) \cdot \beta_{R_iR_jR_k}(t)\ (M^2)^{\alpha_k(o)-1} \cdot \mathrm{Im}\left(-\frac{\tau_i\tau_j\tau_k + e^{-i\pi(\alpha_k(o)-\alpha_i(t)-\alpha_j(t))}}{\sin\pi(\alpha_k(o)-\alpha_i(t)-\alpha_j(t))}\right). \tag{6.29}$$

We have now exhibited the signature factors explicitly, instead of absorbing them into the β's as before. Note the slightly odd structure involving the signature of all these trajectories.[19] For the special case $\alpha_i=\alpha_j$ and α_k the Pomeron, this reduces to what we had previously, with the signatures for α_i and α_j absorbed into β^i_{AC} and β^j_{AC}.

In general, we can say little theoretically about the triple Regge vertex; for the special case where the Reggeon $\alpha_R(t)$ is itself the Pomeron, however, some conclusions can be drawn. This special situation can arise in a reaction like $A+B\rightarrow A+X$, for in this process, the channel $A\bar{A}$ contains the Pomeron. The vertex $\beta_{RRP}(t)$ thus becomes $\beta_{PPP}(t)$, the triple Pomeron vertex.

In the preceding chapter we mentioned the sum rule

$$\langle n\rangle\ \sigma_{AB} = \sum_C \int \frac{d\sigma^C_{AB}}{dtdM^2}\ dtdM^2. \tag{6.30}$$

We note that a lower bound of the right-hand side consists in selecting only the term C=A in the sum over C, and retaining only that part of the integral over the triple Regge region - that is, only the part where t is small and M^2 is large. We then can write the inequality

$$\langle n\rangle\ \sigma_{AB} \geq \int_{M^2_{min}}^{s} dM^2 \int_{t_{max}}^{o} dt\ \frac{d\sigma^A_{AB}}{dtdM^2}\ . \tag{6.31}$$

We are now permitted to replace $d\sigma/dtdM^2$ under the integral by the triple Regge formula: Thus[20]

$$\langle n\rangle\ \sigma_{AB} \geq \int_{M^2_{min}}^{s} dM^2 \int_{t_{max}}^{o} dt\ \frac{1}{16\pi s^2} \left(\frac{s}{M^2}\right)^{2\alpha(t)} \cdot$$

$$\cdot\ (M^2)^{\alpha(o)} |\beta_A(t)|^2 \beta_{PPP}(t)\beta_B, \qquad (6.32)$$

where $\alpha(t)$ stands for the Pomeron.

First suppose $\alpha(t)$ is identically one - a flat Pomeron with unit intercept. Then we find

$$\langle n\rangle\ \sigma_{AB} \geq (\ln s/M^2_{min})\ \frac{1}{16\pi} \int_{t_{max}}^{o} dt\, |\beta_{AA}(t)|^2 \cdot$$

$$\cdot\ \beta_{PPP}(t)\beta_B \sim \ln s. \qquad (6.33)$$

Similarly, if $\alpha(t)=1+\alpha' t$, (i.e., if the Pomeron moves but has unit intercept), we find

$$\langle n\rangle\ \sigma_{AB} \geq \ln \ln s. \qquad (6.34)$$

One may be tempted to say that there is no significant restriction in eqs. (6.33) and (6.34) since they are inequalities for $\langle n\rangle\sigma_{AB}$ and we are already accustomed to a logarithmic increase of this quantity. This, however, is not true since we overlooked an important point - the contribution of the triple Regge formula in the high x regions describes only the production of a single particle.[21)] In fact, kinematics alone (i.e., energy momentum conservation) tells us that in the region $x \geq 0.5$ there can be at most one particle, so that $\langle n \rangle = 1$ in this region. Hence we have actually calculated the contribution of the triple Pomeron region to σ_{AB} itself rather than to $\langle n \rangle \sigma_{AB}$. Mathematically, one can make this statement in a clearer way by considering the momentum conservation sum rule[22)]

$$P_\mu \sigma_{AB} = \sum_C \int dt \int dM^2 \frac{d\sigma^C_{AB}}{dt\,dM^2} q_\mu . \qquad (6.35)$$

Again we bound the right-hand side by C=A and the triple Regge region. We also select the sum of the energy and longitudinal momentum components of the equation. Now we have

$$\sigma_{AB} \geq \int_{M^2_{min}}^{s} dM^2 \int_{t_{max}}^{o} dt \frac{d\sigma^A_{AB}}{dt\,dM^2} (1 - M^2/s) \quad (6.36)$$

and if in this inequality we put $\alpha(t) \equi 1$ we find)

$$\sigma_{AB} \geq \ln(s/M^2_{min}) \frac{1}{16\pi} \int_{t_{max}}^{o} dt |\beta_{AA}(t)|^2 \cdot$$

$$\cdot \beta_{PPP}(t) \beta_B . \qquad (6.37)$$

Now we have derived an inconsistency: We assumed $\alpha(t)=1$, and in particular, therefore, $\alpha(o)=1$; hence total cross sections are constant. Yet (3.25) says total cross sections grow at least logarithmically. The only escape is to assume the triple Pomeron vertex β_{PPP} is identically zero.

A somewhat weaker conclusion obtains if we assume $\alpha(t)=1+\alpha' t$. Then our bound is

$$\sigma_{AB} \geq \frac{\beta_A^2 \beta_{PPP}(o)\beta_B}{16\pi} \cdot \frac{1}{\alpha'} \ln(1+\frac{2\alpha'}{b}\ln s) \quad (6.38)$$

where we have taken the t dependence in the β's to be simply e^{bt}, and now the inconsistency can be removed if we only assume $\beta_{PPP}(o)=0$; the triple Pomeron vertex need only vanish at $t=0^*$.

These results are the first examples of something we'll run into repeatedly later, with more and more devastating consequences; namely, the incompatibility of factorization with a Pomeron which is simply a pole with intercept 1. Note, however, that the fact the Pomeron has been assumed to be a pole here is not the crucial point; even if it were a cut, if it still factorized, we could still be in trouble. For example, if instead of s^α we had $s^\alpha (\ln s)^\nu$, it is easy to see the previous arguments still force $\beta_{PPP}=0$ unless $\nu<-1$; i.e., unless the Pomeron is a soft cut, and thereby yields a total cross-section vanishing faster than $(\ln s)^{-1}$.

What is the experimental situation? Does $\beta_{PPP}(t)$ actually vanish, either at $t=0$ or at all t?

To answer that let us look at the functional form expected from the PPP diagram:

---

* We are reluctant to resolve the problem by letting either β_A or β_B vanish, since this would result in the identical vanishing of the total cross section.

$$\frac{d^2\sigma}{dt\,dx} = \left(\frac{s}{M^2}\right)^{2\alpha_P(t)-1} f_{PPP}(t)$$

$$\approx \left(\frac{1}{1-x}\right)^{1+2\alpha' t} f_{PPP}(t). \qquad (6.39)$$

A nonscaling diffractive contribution can come from a PPR term which would look like

$$\frac{d^2\sigma}{dt\,dx} = \left(\frac{s}{M^2}\right)^{2\alpha_P(t)-1} (M^2)^{\alpha_R(o)-1} f_{PPR}(t)$$

$$\propto \left(\frac{1}{1-x}\right)^{2\alpha_P(t)-\alpha_R(o)} \left(\frac{1}{s}\right)^{1-\alpha_R(o)} \qquad (6.40)$$

which, for fixed s, has even a stronger singularity in 1-x than eq. (6.39). In a complete analysis of data one also has to take into account interference terms as well as lower Regge exchanges such as RRP. Clearly any such exchanges whose $\alpha(t) \leq \frac{1}{2}$ will lead to a vanishing invariant distribution near $x \simeq 1$. Furthermore, the Pomeron singularity, if it is a moving singularity, should, for sufficiently large -t, show a decreasing distribution in 1-x. Recent ISR data for high x (pp,p) reactions are shown in fig. 6.12. A clear peak in the distribution is observed out to the highest measured t value. This substantiates the triple Regge approach and implies the existence of a flat or very slowly moving Pomeron singularity. This is consistent with the results for the Pomeron found in elastic cross

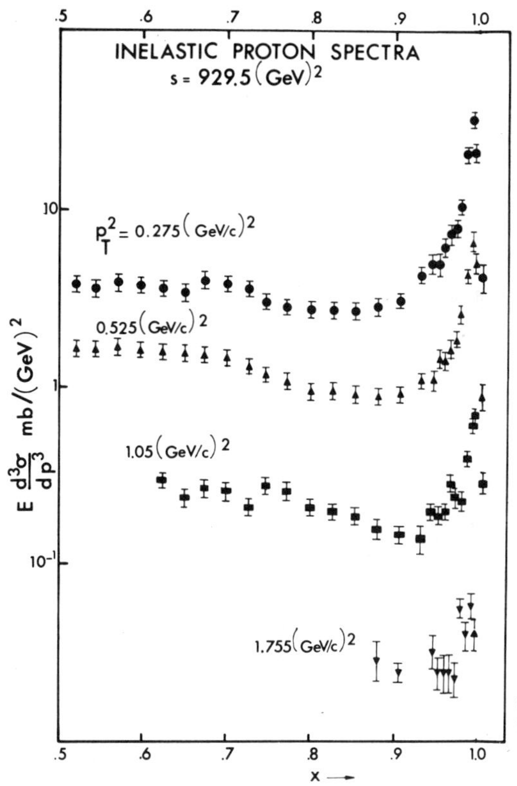

Fig. 6.12. ISR (pp,p) data at high x values. Taken from ref. 23.

section, which were discussed in chapter 4.

The relative amounts of PPP and PPR terms can be found only from a careful analysis of the data at several s values. Recent analyses[24,25] lead to the conclusion that a sizable PPP contribution has to be present.

In fig. 6.13 we show recent ISR data taken at three different energies. Although the error bars are quite big it seems reasonable to conclude that a scaling diffractive term (i.e., PPP) is present. The next interesting question is then its t-dependence. We saw above that a Regge pole model calls for a vanishing residue at t=0 in order to avoid the appearance of an increasing cross

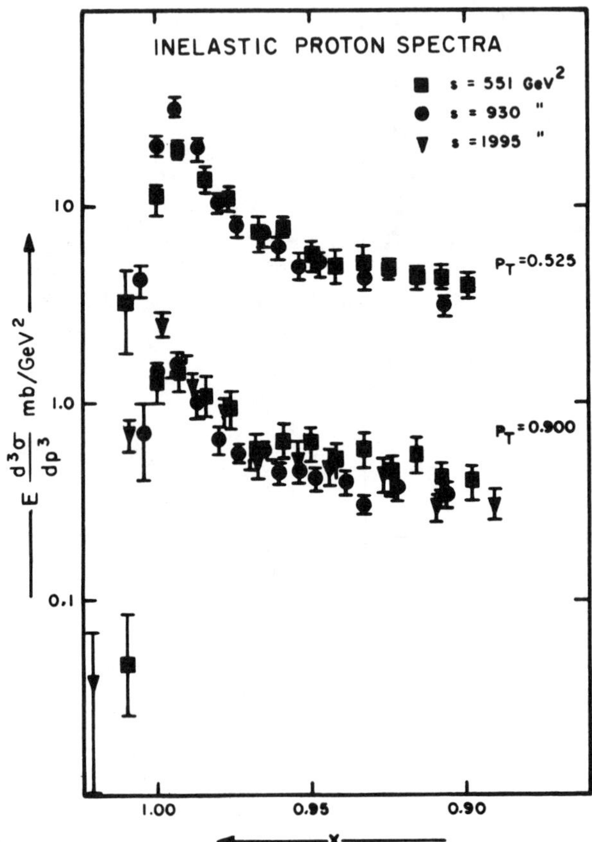

Fig. 6.13. ISR (pp,p) data. Comparison between different s-values. Taken from ref. 26.

section. No such decrease is evident in fig. 6.14 which shows the t dependence of the same data plotted before in fig. 6.12. This does not rule out the possibility that a decrease may exist at much lower $|t|$ values. In any case it will presumably be difficult to disentangle the t-dependence of the different components. We note that the t-dependences shown in fig. 6.14 are generally steeper than those encountered in diffractive resonance production in chapter 4. The latter can be connected via standard duality arguments to a PPR

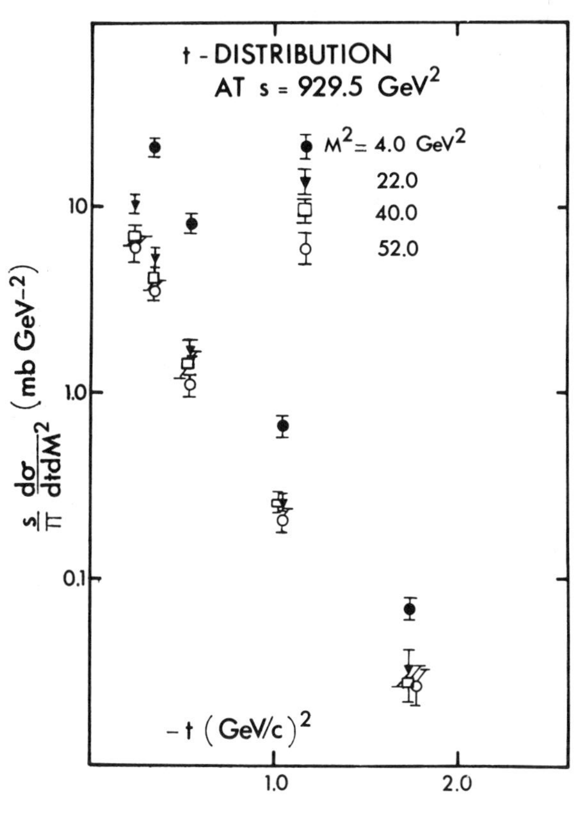

Fig. 6.14. Same data as in fig. 6.12 plotted vs. t. Taken from ref. 23.

term. This method was used in ref. 25 to conclude that a dominant PPP mode must be present.

Since the data are consistent with an almost flat (in t) PPP vertex we may conclude that they presumably contribute to a logarithmic increase in the total cross section. It will be interesting to have a clear cut experimental determination of this as well as other aspects of diffractive production.

Finally, let us remark that a triple-Regge analysis should in principle also apply to non-diffractive processes with another Regge trajectory replacing the Pomeron. Attempts to use this description for particle spectra at low s values have not been too successful.[21] The principal reason is presumably the fact that such production data are not available at high x values (e.g., $x>0.9$) and the triple Regge representation may very well fail below $x \approx 0.9$. It is difficult to determine a-priori what is the expected range of x in which a triple Regge representation holds. The analysis of pp production data at low s values in ref. 27 covered regions of $x \lesssim 0.8$ and showed that this representation calls for very low intercepts of the exchanged baryon trajectories (like -1 or -2). Nevertheless, the general hierarchy of Regge exchanges, namely the order mesonic> baryonic>exotic α values, seems to hold. We have therefore an indication that an exchange model can be a successful candidate for the description of the data, but we have to await detailed measurements at the NAL and ISR to determine its exact properties.

7

Two-Particle Inclusive Distributions

The situation here is analogous to what happens in the one-particle inclusive case, so we shall content ourselves with a few remarks of interest and not attempt an exhaustive discussion.

We wish to describe the reaction

$$A + B \to C + D + X$$

and in particular to characterize the two-particle distribution function ρ_{AB}^{CD}. To do this it is, as before, convenient to employ the generalized optical theorem, which states that

$$\rho_{AB}^{CD} = \frac{1}{s} A_{A\bar{C}\bar{D}B\to A\bar{C}\bar{D}B}\Bigg|_{\text{forward}} \tag{7.1}$$

As an illustration, let us look at the kinematic region where $s_{12}=(q_1+q_2)^2$ is large, but where $t_1=(p_1-q_1)^2$ and $u_1=(p_2-q_2)^2$ are small. Thus particle C (momentum q_1) continues on in the direction of A, while D (momentum q_2) is produced in the direction of B. The Regge picture of this limit is sketched in fig. 7.1:

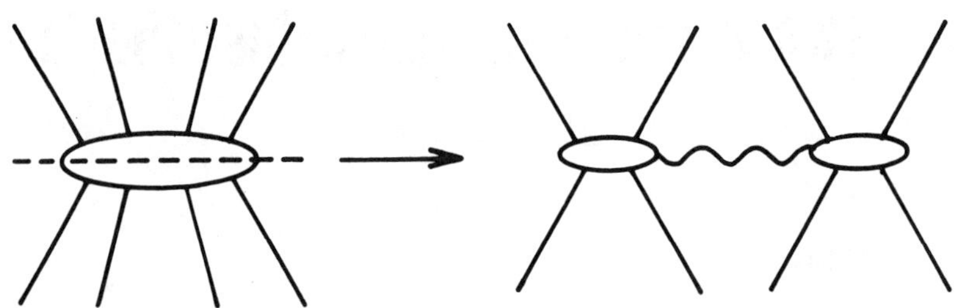

Fig. 7.1

Evidently, we expect that

$$\rho_{AB}^{CD} \to \beta_{AC}(q_{1T}, \frac{Y}{2}-y_1)\ \beta_{BD}(q_{2T}, \frac{Y}{2}-y)$$
$$+ \sum_i \beta_{AC}^i(q_{1T}, \frac{Y}{2}-y_1)\ s^{\alpha_i(o)-1}\ \cdot$$
$$\cdot\ \beta_{BD}^i(q_{2T}, \frac{Y}{2}+y_2)\ \tau_i \qquad (7.2)$$

where we have written both the leading contribution from the Pomeron and the contribution of lower lying trajectories $\alpha_i(t)$. The functions β_{AC}, etc., are evidently precisely the same functions we met with in the one-particle inclusive situation. Therefore, we can also calculate the two-particle correlation function, and we find that the leading terms cancel:

$$f_{AB}^{CD} = \frac{1}{\sigma_{AB}}\ \rho_{AB}^{CD} - \frac{1}{\sigma_{AB}^2}\rho_{AB}^{C}\rho_{AB}^{D} \propto e^{-Y/L} \qquad (7.3)$$

where we use the correlation length $L=1/\left[1-\alpha_i(o)\right]\sim 2$. The Regge picture is thus automatically a weakly correlated picture.[28)]

As a matter of fact, detailed calculations can lead to further cancellations.[14)] Using the exchange degenerate pattern discussed in the previous chapter, we find from eqs.(6.19) and (6.20) that

$$\rho^{\pi^-\pi^-}_{p\pi^+} = \beta_{p\pi^-}\ \beta_{\pi^+\pi^-} + \left(\frac{s}{s_o}\right)^{-1/2}\left[\beta^f_{p\pi^-}\ \beta^f_{\pi^+\pi^-} - \beta^\rho_{p\pi^-}\cdot\beta^\rho_{\pi^+\pi^-}\right] + O(s^{-1}) = \beta_{p\pi^-}\beta_{\pi^+\pi^-} + O(s^{-1}). \tag{7.4}$$

Similarly, one is led to predict early scaling for (pp, $\pi^-\pi^-$), (K^+p, $\pi^-\pi^-$), (pp, $\pi^+\pi^-$) and (K^+p, $\pi^-\pi^+$) in the double fragmentation region. Some evidence exists at low s values supporting early scaling in the first two reactions. In changing particles A and C into $\overline{A}$ and $\overline{C}$ one has to change the sign of the ρ and ω couplings and therefore the above mentioned cancellation will no longer occur. We expect, then, that the double fragmentation in reactions like ($\bar{p}$ p, $\pi^+\pi^-$) and (K^-p, $\pi^+\pi^-$) will show the $s^{-1/2}$ trend of approach to scaling.

At the edges of the double fragmentation region we recover triple Regge behavior analogous to that discussed before. For example, suppose we hold $t_1=(p_1-q_1)^2$, the momentum transfer from A to C, fixed and let $s/\overline{s}_1\to\infty$ where $\overline{s}_1=(P-q_1)^2$ is the missing mass associated with D and X. Then conventional two-body Reggeism suggests that

$$\rho_{AB}^{CD} \sim \frac{1}{s} \cdot \left(\frac{s}{\bar{s}_1}\right)^{2\alpha_1(t_1)} \tag{7.5}$$

where α_1 is the leading trajectory in the $A\bar{C}$ channel.

Next let us hold $t_2 = (p_1 - q_1 - q_2)^2$ fixed and let $\bar{s}_1/M^2$, where $M^2 = (p - q_1 - q_2)^2$ is the missing mass, grow. Then we should expect

$$\rho_{AB}^{CD} \sim \frac{1}{s} \cdot \left(\frac{s}{\bar{s}_1}\right)^{2\alpha_1(t_1)} \cdot \left(\frac{\bar{s}_1}{M^2}\right)^{2\alpha_2(t_2)} \tag{7.6}$$

where α_2 is the leading trajectory in the $A\bar{C}\bar{D}$ channel. Finally, we may let $M^2 \to \infty$. The successive limits are depicted in fig. 7.2, and the resulting form for ρ is

$$\rho_{AB}^{CD} \to \frac{\beta_B}{s} \left| \beta_{AC_1}(t_1) \left(\frac{s}{\bar{s}_1}\right)^{\alpha_1(t_1)} \cdot \right. \\ \left. \cdot \beta_{R_1 R_2}^{D}\left(t_1, \frac{s_{12}\bar{s}_1}{s}, t_2\right) \left(\frac{\bar{s}_1}{M^2}\right)^{\alpha_2(t_2)} \right|^2 \cdot \\ \cdot\ \beta_{R_1 R_2 P}(t_2) \cdot M^2 . \tag{7.7}$$

Here $\beta_{R_1 R_2}^{D}$ is the double Regge vertex which couples D to the Reggeon α_1 and the Reggeon α_2. $\beta_{R_1 R_2 P}$ is the triple Regge vertex connecting the Pomeron and α_1 and α_2. The other symbols we have met with previously. We have indicated that the double Regge vertex depends not only on the momentum transfers t_1 and t_2 which feed into it, but also, in general, on the combination $K = s_{12}\bar{s}_1/s$.

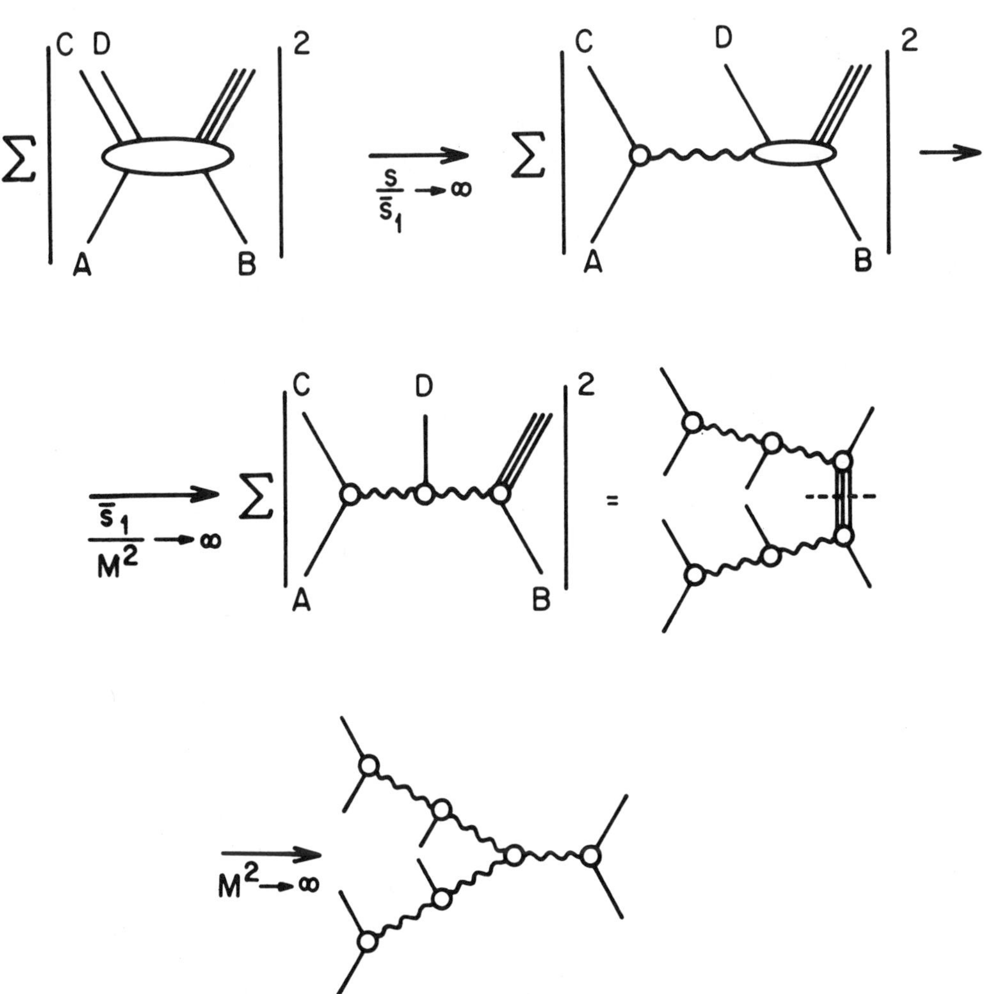

Fig. 7.2

We shall defer discussion of the reason for this to the chapter on the multi-Regge model; at present we need to make no use of this fact.

The existence of this complicated form in a particular kinematic region permits us to obtain more stringent lower bounds on total cross-sections by making use of sum rules, as in the previous section. In particular, we recall the formula[22]

$$(P-q_1)_\mu \rho^{C}_{AB} = \sum_D \int \frac{d^3 q_2}{\omega} (q_2)_\mu \rho^{CD}_{AB}. \qquad (7.8)$$

We add the longitudinal and time components of this equation, select D to be such that α_1 can be the Pomeron, and keep only the region of the integral corresponding to eq. (7.7). This clearly provides a lower bound to the right-hand side, and the resulting inequality[20] reads

$$\beta_{PPP}(t_1) \geq \frac{1}{\bar{s}_1} \int \left| \beta^{D}_{PR_2} (t_1, K, t_2) \left(\frac{\bar{s}_1}{M^2}\right)^{\alpha_2(t_2)} \right|^2 \cdot \beta_{R_1R_2P}(t_2) M^2 \frac{2\omega_2}{\sqrt{s}-2\omega_1} \frac{d^3 q_2}{\omega_2}. \qquad (7.9)$$

Now let $t_1 \to 0$. We will show later that in this limit $\beta^{D}_{PR_2}$ becomes a function of t_2 only. We concluded before that as $t_1 \to 0$, $\beta_{PPP}(t_1) \to 0$; hence the right-hand side must vanish as well. This requires that

$$\beta_{R_2R_2P}(t_2) \to 0 \qquad (7.10)$$

or $$\beta^{D}_{PR_2} (t_2) \to 0. \qquad (7.11)$$

Thus either the Pomeron-Reggeon-particle vertex or the Pomeron-Reggeon-Reggeon vertex must vanish.

But either of these two alternatives is

disastrous. For if the Pomeron-Reggeon-Particle vertex vanishes, then let us continue t_2 to a physical particle mass,[29] in which event we learn that the Pomeron-particle-particle vertex vanishes - and the Pomeron decouples from all particles. The same result applies if we say $\beta_{PR_2R_1}(t_2)=0$, for again we continue t_2 to the mass of a particle on α_2; again the Pomeron decouples.

We are now in even more serious straits than before; it appears that we simply cannot maintain both factorization and a Pomeron intercept of unity. We are compelled to say one of the following:

(i) The Pomeron intercept is slightly less than unity:[39] $\alpha(o)=1-\varepsilon$. This means that cross sections vanish like powers of s: $\sigma \to s^{-\varepsilon}$. Experimentally, there is no evidence for such behaviors in pp scattering through the ISR energy range. Hence ε must be exceedingly small, and it is obviously embarrassing to have to appeal to the existence of a very small number without any theoretical understanding of why it should be so small; small numbers rarely happen by accident.

(ii) The Pomeron intercept is at one, but the continuation of the vertex functions from $t<0$ to the particle mass fails. The reason for this failure may lie in the existence of cuts associated with the Reggeon; such cuts are believed to exist for other reasons anyway.

(iii) The Pomeron is not a pole at all, but is

some collection of branch points. In this event factorization, no doubt, fails, and the entire discussion given above, which led us into difficulties, becomes irrelevant. This would mean that our simple Regge mnemonic, which has, as we have seen, had some success, cannot be exactly right. But it may well be approximately right, and its partial failure may still be enough to escape from our dilemma.

Up to here we have discussed distributions in the double fragmentation region. The correlation structure mentioned in chapter 3 appears in the distributions in the central region. There we look at diagrams of the type shown in fig. 7.3.

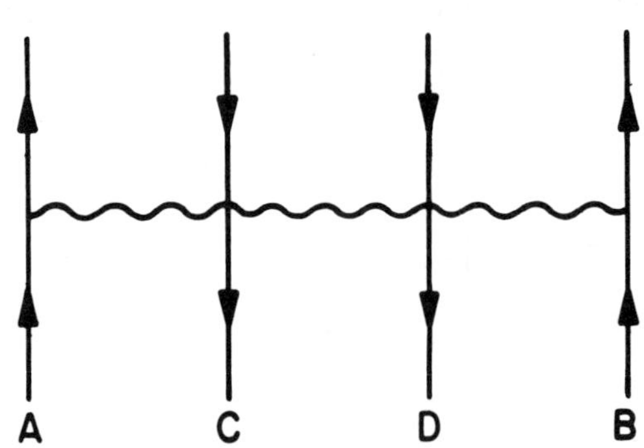

Fig. 7.3

The characteristic matrix element will be

$$
\begin{aligned}
\rho_{AB}^{CD} = {} & \beta_A \beta_C (q_{CT}) \beta_D (q_{DT}) \beta_B \\
& + \sum_i \beta_A \beta_C^{iP} (q_{CT}) \tau_i \beta_D^{iP} (q_{DT}) \beta_B (s_{CD})^{\alpha_i(o)-1} \\
& + \sum_i \beta_A^i \beta_C^{iP} (q_{CT}) \tau_i \beta_D (q_{DT}) \beta_B (s_{AC})^{\alpha_i(o)-1} \\
& + \sum_i \beta_A \beta_C (q_{CT}) \tau_i \beta_D^{iP} (q_{DT}) \beta_B^i (s_{BD})^{\alpha_i(o)-1} \\
& + \ldots
\end{aligned}
\qquad (7.12)
$$

where we have written the contributions of a PPP term as well as of PRP, RPP, and PPR terms. If both C and D are in the central region we note that PPP and PRP are scaling contributions and RPP, PPR, etc., are corrections to scaling.[12)] Limiting ourselves to the terms that survive asymptotically we arrive at

$$f_{AB}^{CD} = \frac{1}{\sigma_{AB}} \rho_{AB}^{CD} - \frac{1}{\sigma_{AB}^{2}} \rho_{AB}^{C} \rho_{AB}^{D}$$

$$= \sum_{i} \beta_C^{iP} (q_{TC}) \tau_i \beta_D^{iP} (q_{TD}) \cdot (s_{CD})^{\alpha_i(o)-1} \tag{7.13}$$

which shows the expected short-range correlation since

$$(s_{CD})^{\alpha_i(o)-1} \sim \exp(-|y_C - y_D|/L),$$

where $L = \left[1-\alpha_i(o)\right]^{-1} \simeq 2.$

We note that f_{AB}^{CD} is independent of AB. This again is a consequence of the factorization of the Pomeron residues. The terms β_C^{iP} denote a PRCC coupling. It is the same vertex that dominates corrections to scaling in the central region of the one-particle distribution which we saw in eq. (6.21). Thus the correlations in the central region can be connected to the energy dependence of the approach to scaling of the one-particle inclusive distribution in the central region.

Let us use this formalism to derive some simple results that have some bearing on the experimental data which were discussed in chapter 3. Since the particles produced in the central region are mainly pions we shall consider only them. In terms of the leading Regge pole approximation used before we will find that

$$f^{\pi^-\pi^-} = f^{\pi^+\pi^+} = (s_{\pi\pi})^{\alpha(o)-1} \left[(\beta_{\pi^+}^{fP})^2 - (\beta_{\pi^+}^{\rho P})^2\right]$$

$$f^{\pi^o\pi^o} = (s_{\pi\pi})^{\alpha(o)-1} (\beta_{\pi^+}^{fP})^2$$

$$f^{\pi^+\pi^o} = f^{\pi^-\pi^o} = (s_{\pi\pi})^{\alpha(o)-1} (\beta_{\pi^+}^{fP})^2$$

$$f^{\pi^-\pi^+} = f^{\pi^+\pi^-} = (s_{\pi\pi})^{\alpha(o)-1} \left[(\beta_{\pi^+}^{fP})^2 + (\beta_{\pi^+}^{\rho P})^2\right] \tag{7.14}$$

where we have used $\beta_{\pi^+}^{fP} = \beta_{\pi^-}^{fP} = \beta_{\pi^o}^{fP}$ and $\beta_{\pi^+}^{\rho P} = -\beta_{\pi^-}^{\rho P}$. It is now evident that

$$R^{ch,ch} = \sigma \frac{f^{\pi^-\pi^+} + f^{\pi^+\pi^+} + f^{\pi^+\pi^-} + f^{\pi^-\pi^-}}{\rho^{\pi^-}\rho^{\pi^+} + \rho^{\pi^+}\rho^{\pi^+} + \rho^{\pi^+}\rho^{\pi^-} + \rho^{\pi^-}\rho^{\pi^-}}$$

$$= R^{o,o} = R^{ch,o} \tag{7.15}$$

which is consistent with the experimental results displayed in fig. 3.1. This is actually just a statement of charge independence in the central region. We see that the two-particle distribution, when analyzed in terms of Regge poles, has a structure which is very similar to that of elastic cross-sections. Drawing on this analogy we may invoke once again the principle of duality[3) which would suggest that the leading term in $f^{\pi^-\pi^-} = f^{\pi^+\pi^+}$ vanishes because these Regge exchanges are dual to exotic s-channels. It is clearly very reasonable to assume that $f^{\pi^-\pi^+} > f^{\pi^-\pi^-}$ because of resonance production. It remains, however, to be seen if the short-range component of $f^{\pi^-\pi^-}$ indeed vanishes.

It seems fairly safe to assume that the description of the short-range behavior will roughly follow the simple Regge picture. However, long-range effects certainly exist. Such terms can for instance arise if the Pomeron is a cut instead of a moving pole. This will introduce powers of $y_C - y_A$, $y_D - y_C$, $y_B - y_D$, as well as of all other rapidity differences in the distribution. Such terms will not cancel out in eq. (7.13). Nevertheless, their modification of f_{AB}^{CD} will presumably have a soft rapidity dependence instead of the characteristic exponential damping of a PRP term. We may, therefore, hope to be able to separate and identify the PRP term even in the presence of long-range correlations.

8

Several Particle Exclusive Processes: Multi-Regge Exchange

The reaction $A+B \to C_1+C_2+C_3$, with momenta $p_1+p_2=q_1+q_2+q_3$, depends on five independent scalar variables. A number of useful variables are the following:

$$\begin{aligned}
s &= (p_1 + p_2)^2 \\
t_1 &= (p_1 - q_1)^2 \\
t_2 &= (p_1 - q_1 - q_2)^2 \\
s_{12} &= (q_1 + q_2)^2 \\
s_{23} &= (q_2 + q_3)^2 \\
s_{13} &= (q_1 + q_3)^2 \\
K_{13} &= s_{12}s_{23}/s_{13} = K \\
K_{23} &= s_{21}s_{13}/s_{23} \\
K_{12} &= s_{13}s_{32}/s_{12} \ .
\end{aligned} \tag{8.1}$$

Obviously not all of these are independent; a convenient set of 5 independent ones is s_{12}, s_{23}, t_1, t_2, and K. These are indicated in fig. 8.1.

Now we know that if we continue to $t_2=m^2$, where m is the mass of some particle in the $B\overline{C}_3$

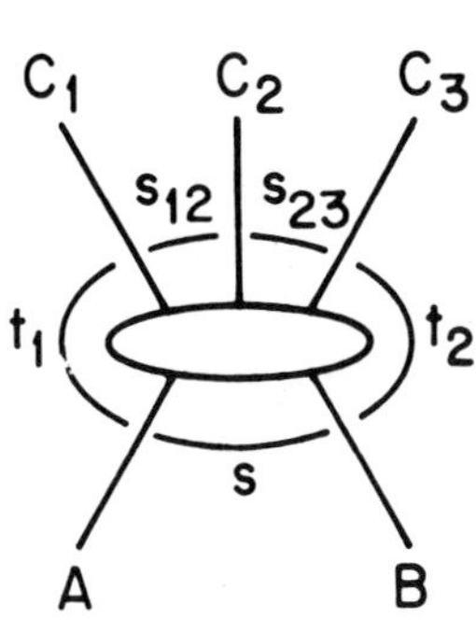

Fig. 8.1

channel, then there will be a pole in t_2 at m^2, and the residue of this pole must be the physical two-body scattering process $A(B\bar{C}_3)\rightarrow C_1C_2$ (the particle of mass m is denoted $(B\bar{C}_3)$) times the coupling constant of $(B\bar{C}_3)$ to B and C_3. When s_{12} becomes large, with t_1 fixed, the two-body process has Regge behavior, and is proportional to $(s_{12})^{\alpha_1(t_1)}$, where α_1 is the leading trajectory in the $A\bar{C}_1$ channel.

We may hope that this behavior continues to be true even if we continue t_2 from its mass shell value of m^2 down to some negative value.*) If it does, and if we make the same argument on t_1, then it is very strongly suggested that in the double limit $s_{12}\rightarrow\infty$, t_1 fixed, and $s_{23}\rightarrow\infty$, t_2 fixed, we should have[31)]

$$T_{AB\rightarrow C_1C_2C_3} \longrightarrow \beta_{AC_1}(t_1)\left(\frac{s_{12}}{s_0}\right)^{\alpha_1(t_1)} \cdot$$
$$\cdot R_{C_2}(t_1,K,t_2)\left(\frac{s_{23}}{s_0}\right)^{\alpha_2(t_2)} \beta_{BC_3}(t_2). \qquad (8.2)$$

* Indulging freely in such continuations, we recall, could be what led us into trouble before, at the end of chapter 7. It is, therefore, necessary to maintain some suspicion about this argument, although it is consistent with the pure Regge pole picture we are discussing here.

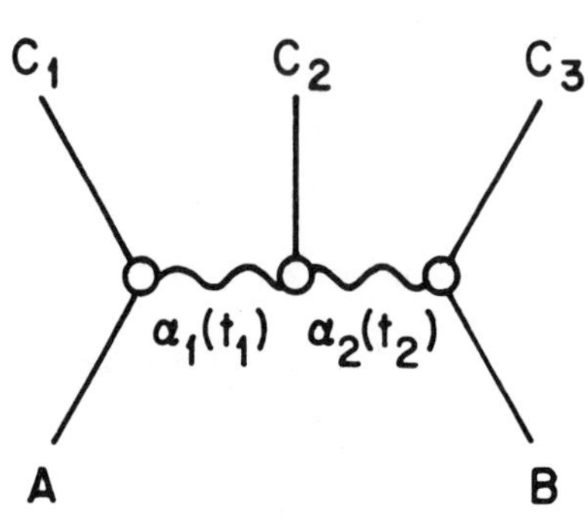

Fig. 8.2

In pictures, this formula is displayed in fig. 8.2. β_{AC_1} and β_{BC_3} are normal Regge residue functions; $R_{C_2}(t_1,K,t_2)$ is the coupling of C_2 to the Reggeons α_1 and α_2; this is the same vertex we ran into in chapter 7. We know that the β's depend only on t_1 and t_2; hence, insofar as the entire reaction depends on K that dependence must be in R. The signature factors have also been absorbed into R.

The detailed form of R is a bit subtle. It turns out that it may be written as follows:[32)]

$$R = (1+\tau_1 e^{-i\pi\alpha_1})(\tau_1\tau_2+e^{i\pi(\alpha_1-\alpha_2)})V_1(K)K^{-\alpha_1}$$
$$+(1+\tau_2 e^{-i\pi\alpha_2})(\tau_1\tau_2+e^{i\pi(\alpha_2-\alpha_1)})V_2(K)K^{-\alpha_2} . \tag{8.3}$$

The reason for this form is that we must guarantee that the final amplitude does not have simultaneous discontinuities in s_1 and s_2; there must be no overlapping singularities.[33)] That this goal is accomplished by the form given is clear if we write out the entire expression for T:

$$T_{AB\to C_1C_2C_3} \to \beta_{AC_1} s_{12}^{\alpha_1} \xi_1 \eta \left(\frac{s}{s_{12}s_{23}}\right)^{\alpha_1} V_1 \cdot$$

$$\cdot\ s_{23}^{\alpha_2} \beta_{BC_3} + \beta_{AC_1} s_{12}^{\alpha_1} \xi_2 \eta \left(\frac{s}{s_{12}s_{23}}\right)^{\alpha_2} \cdot$$

$$\cdot\ s_{23}^{\alpha_2} \beta_{BC_3} . \tag{8.4}$$

(We have set $s_0=1$, $\xi=1+\tau e^{-i\pi\alpha}$ is the usual signature factor, and $\eta=\tau_1\tau_2+e^{+i\pi(\alpha_1-\alpha_2)}$.) The first term contains $(\frac{s}{s_{23}})^{\alpha_1}$ $(s_{23})^{\alpha_2}$ while the second term contains $(s_{12})^{\alpha_1}$ $(s/s_{12})^{\alpha_2}$. Thus simultaneous singularities occur only in s and s_{23}, or in s and s_{12}, but not in s_{12} and s_{23}. The appearance of ratios like (s/s_{23}) is not surprising when we go back to our original heuristic argument which suggested a double Regge form in the first place. We could, for example, just as well have thought of the process as $AB\to C_1(C_2C_3)$ instead of as $A(B\overline{C}_3)\to C_1C_2$. Then from this point of view Reggeism would have suggested the behavior $(s/s_{23})^{\alpha_1}$ instead of $(s_{12})^{\alpha_1}$. Obviously, insofar as we really are in the Regge region, s/s_{23} and s_{12} are proportional, so from simply a power dependence point of view it doesn't matter which we use. The overlapping singularity argument, however, does decide the matter, and the appropriate form is as given above.

What kinematic region does the double Regge limit correspond to? That is, where can we expect this double Regge formula to hold? We have

$$s_{12} = (q_1 + q_2)^2$$

$$= m_{C_1}^{\ 2} + m_{C_2}^{\ 2} + 2\omega_1\omega_2 - 2q_{1L}q_{2L} - 2\vec{q}_{1T}\vec{q}_{2T}. \quad (8.5)$$

In order for this to be large, yet for t to be small, and hence for $\vec{q}_T$ to be small, we evidently must have $\omega_1 >> \omega_2$. Similarly, s_{23} is large if $\omega_3 >> \omega_2$. Thus the kinematic region of interest is where the two outside particles are fast; $\omega_1 \approx \omega_3 \approx \sqrt{s}/2$, while ω_2 is small. This arrangement is known as strong ordering. We note that in terms of rapidities

$$s_{12} = m_{C_1}^{\ 2} + m_{C_2}^{\ 2} + 2m_{T_1}m_{T_2} \cosh(y_1 - y_2) \qquad (8.6)$$

so that s_{12} is large if $y_1 - y_2$ is large. We also see that when this is the case

$$s_{12}s_{23} \approx m_{T_1}m_{T_3}m_{T_2}^{\ 2}\, e^{y_1 - y_3} \approx s\, m_{T_2}^{\ 2} \qquad (8.7)$$

so that $s_{12}\, s_{23}$ is of order s. Hence $s >> s_{12}$ and $s >> s_{23}$. Note also that s_{13} is of order s.

The picture is, therefore, that C_1 comes out along A and is fast; C_3 comes out along B and is fast; and C_2 hangs around in the center doing whatever it wants to do. Using our earlier terminology, C_1 and C_3 are in the fragmentation regions of A and B while C_2 is in the central region.

Finally, we note that

$$K = (m_{C_2}^{\;2} + q_{2T}^{\;2}) = m_{T_2}^{\;2} \tag{8.8}$$

so this variable describes the detailed motion of the central, slow particle; it is, therefore, evidently not of great phenomenological interest. In fact, if, say, $t_1=0$, then q_{2T} depends only on t_2, so that K is a function of t_2. That means the K dependence of the amplitude disappears if either t is zero.[34] Since we expect the whole picture to make sense only if the t's are small, it is likely that the K dependence is weak.

A particular reaction can have the double Regge form in several regions. As an example, let us look at the process $\pi^- p \to \pi^+ \pi^- n$. We distinguish three regions of the Dalitz plot, as follows:

(i) $s \sim s_{\pi^- n} >> s_{\pi^- \pi^+}, s_{\pi^+ n} >>$ masses. There are now two subregions. In one, we have $t_{\pi^- \pi^-}$ and t_{pn} small; in the other $t_{\pi^- n}$ and $t_{p\pi^-}$ are small. The two possibilities are displayed in fig. 8.3.

The leading Regge trajectories are easily read off from the figure. In case (a) we have

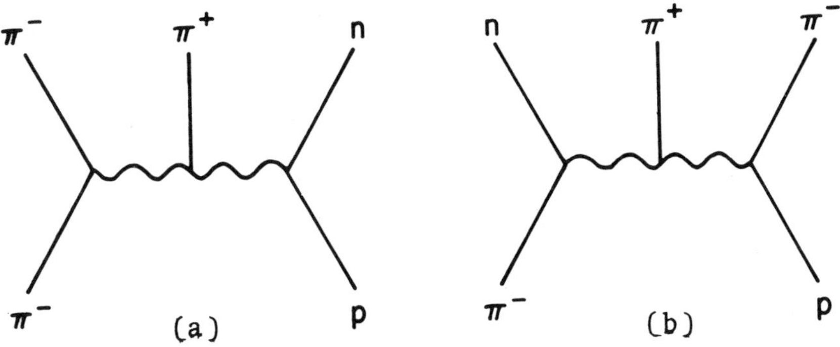

Fig. 8.3

α_1=Pomeron and $\alpha_2=\omega$ or π, or $\alpha_1=\rho$ and $\alpha_2=\omega$ or π. In case (b), the dominant trajectories are α_1=nucleon and α_2=nucleon.

(ii) $s \sim s_{\pi^+n} \gg s_{\pi^+\pi^-}, s_{\pi^-n} \gg$masses. The two subregions now have (a) $t_{\pi^-\pi^+}$ and t_{pn} small, in which case the leading trajectories are α_1=exotic, $\alpha_2=\pi$ or ρ or (b) t_{π^-n} and $t_{p\pi^+}$ small, with α_1=nucleon and α_2=nucleon. The pictures are in fig. 8.4.

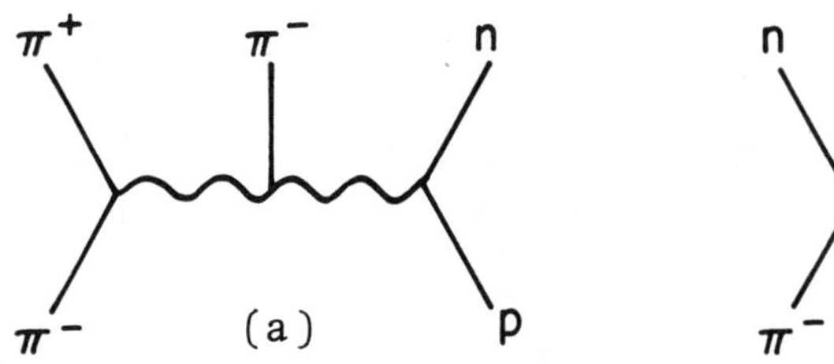

Fig. 8.4

(iii) $s \sim s_{\pi^+\pi^-} \gg s_{\pi^+n}, s_{\pi^-n} \gg$masses. Now in case (a), $t_{\pi^-\pi^-}$ and $t_{p\pi^+}$ are small, so that α_1=Pomeron and α_2=nucleon while in case (b) $t_{\pi^-\pi^+}$ and $t_{p\pi^-}$ are small, and α_1=exotic, $\alpha_2=\Delta$. Fig. 8.5 shows the situation.

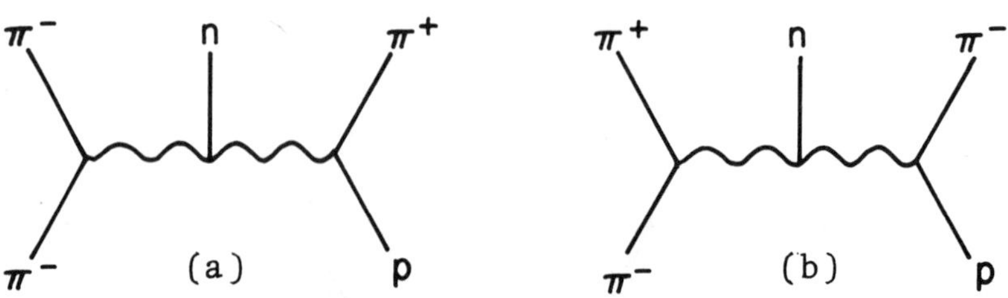

Fig. 8.5

The three main regions are contained in the Dalitz plot shown in fig. 8.6.

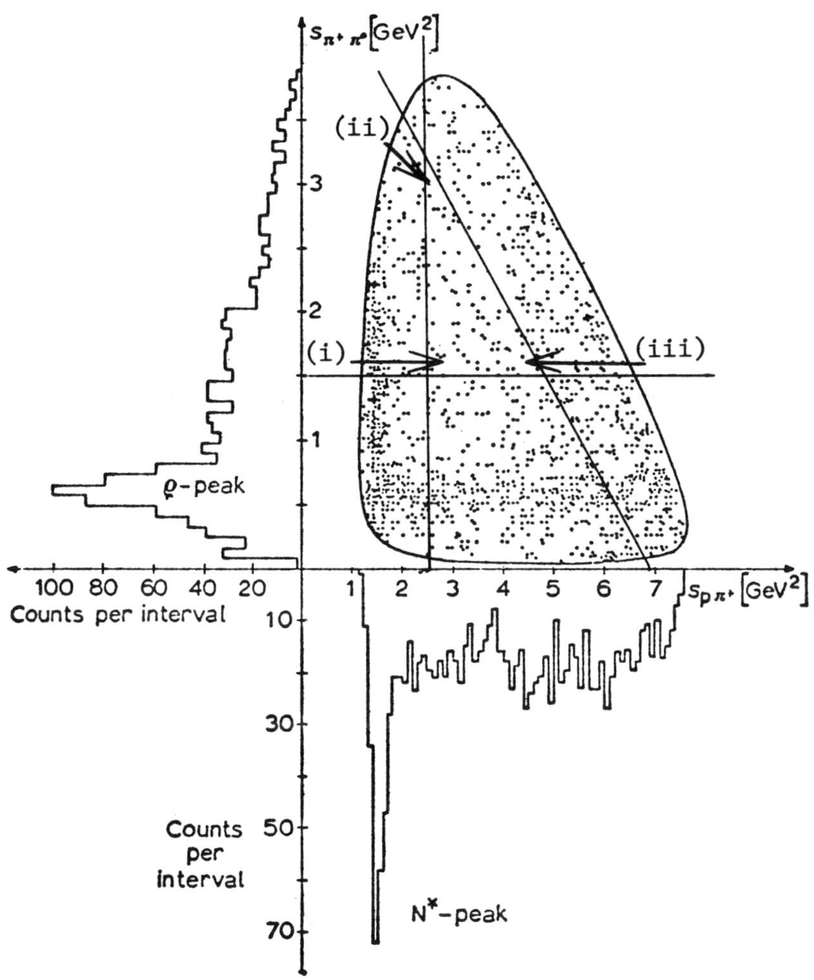

Fig. 8.6. Dalitz plot for the reaction $\pi^+p \to \pi^0\pi^+p$ at 4 GeV. The double Regge regions are included in the corners of the triangle inside the plot. However, not all events in the corners are double Regge events; one must select only those events with small momentum transfers. We do not mean to imply that 4 GeV is high enough energy to justify a double Regge analysis. The Dalitz plot is taken from ref. 35.

The double-Regge regions occupy the corners of a triangle in the center of the Dalitz plot. The position of the triangle is determined by the requirement that s_{12} and s_{23} must be large in the Regge region. The corners are determined by the strong ordering: $s_{13} \sim s >> s_{12}, s_{23}$. In each corner, the two subregions are separated by selecting events in which one pair of t's or the other is small. In this way, double-Regge events can be selected and compared with (8.2). Double-Regge events can equally well be selected by making a rapidity plot rather than a Dalitz plot. If one simply displays all small q_T events by plotting the position of y_1, y_2, and y_3 on a scale stretching from $-\frac{Y}{2}$ to $\frac{Y}{2}$, then events with large gaps between y_1 and y_2, and between y_2 and y_3 are the desired ones.

The characteristic behavior to be expected of double-Regge events is a rapid falloff in t_1 and t_2 - an exponential falloff like that seen in two-body exclusive reactions - and a power dependence in s_{12} and s_{23} with a power characteristic of α_1 and α_2.

Little detailed data fitting with the double-Regge formula has been done in which s_{12} and s_{23} are really large. Let us describe briefly one fit, to 25 GeV $\pi^- p \to \pi^+ \pi^- p$ events.[36] Events in which one $\pi^+\pi^-$ pair has low mass are selected. Call this pair X. Then cuts are made to retain only events in which $|t_{\pi\pi} + 2t_{pp}| \leq 0.8$ (GeV)^2, and in which $s_{\pi^-X}>2\ (\text{GeV})^2$ and $s_{pX}>4\ (\text{GeV})^2$. This eliminates 90% of all events. Thus double-Regge events constitute a distinct minority of everything that happens. The picture which should describe the process in this region is shown in fig. 8.7, and the expected form of the amplitude is

$$T = \beta_{\pi^-}(t_{\pi^-})(s_{\pi^- X})^{\alpha_1(t_{\pi^-})} R(t_{\pi^-}, K, t_p) \cdot$$

$$\cdot\ (s_{Xp})^{\alpha_2(t_p)}\ \beta(t_p). \qquad (8.9)$$

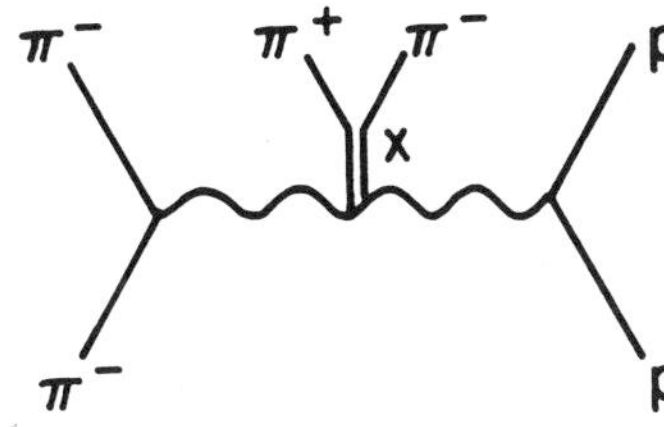

Fig. 8.7

Depending on what mass we select for X, α_1, and α_2 can be different things. If $m_X{}^2=m_\rho{}^2$, we would expect α_1=Pomeron and $\alpha_2=\rho$. If $m_X{}^2=m_{f_o}{}^2$, we would expect both α_1 and α_2 to be Pomerons.

It turns out, as shown in fig. 8.8, that there is no sign of f_o production. The data thereby suggest the absence of double-Pomeron exchange, and the vanishing of $R_{PP\ fo}$. There is, however, a large ρ peak. Selecting only events in this peak, best fits are achieved with $\alpha_1 \sim 1$ and $\alpha_2 \sim \frac{1}{2}$, as shown in figure 8.9. Thus, the identification of α_1 as the Pomeron and α_2 as ρ for ρ production does seem to be confirmed -

albeit within rather broad errors.

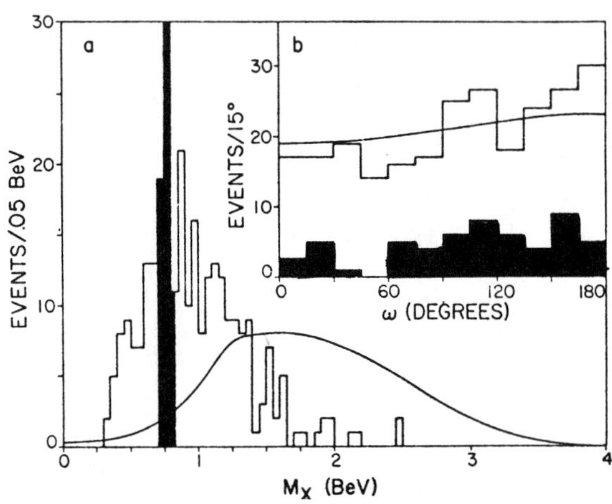

Fig. 8.8. Number of events as a function of the mass of X. The ρ peak is shown in black. There is no sign of an fo peak. The insert in the figure shows the insensitivity of the data to the angle ω, which is related to K. Taken from ref. 36.

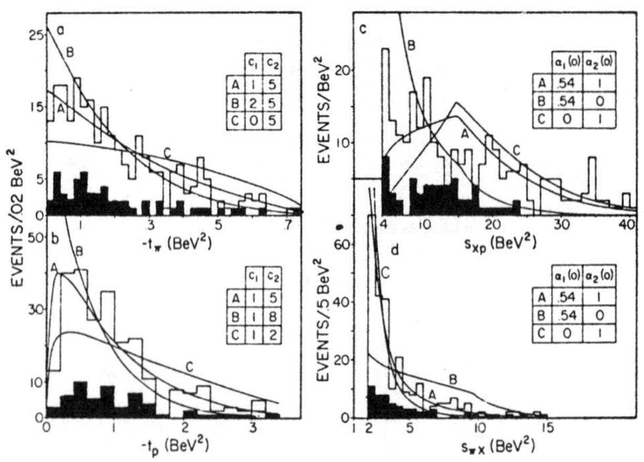

Fig. 8.9. Number of events versus t_π, t_p, s_{Xp} and $s_{\pi X}$. The solid curves are fits with the indicated values for $\alpha_1(o)$ and $\alpha_2(o)$. The black region represents events in the ρ peak. Taken from ref. 36.

The extension of our discussion from the process $A+B\to C_1+C_2+C_3$ to the general n-body exclusive case, namely the reaction $A+B\to C_1+C_2+..+C_n$ with momenta $p_1+p_2\to q_1+q_2...+q_n$ is reasonably straightforward. There are altogether 3n-4 independent variables on which the amplitude depends. In analogy to the n=3 case we may define a number of useful invariants:

$$
\begin{aligned}
s &= (p_1+p_2)^2 \\
s_{i,i+1} &= (q_i+q_{i+1})^2 \qquad \text{(subenergies)} \\
t_i &= (p_1-q_1-\cdots\cdots-q_i)^2 = (p_2-q_{i+1}-\cdots-q_n)^2 \\
& \qquad\qquad \text{(momentum transfers)} \\
s_i &= (q_1+\cdots\cdots+q_i)^2 \\
\bar{s}_i &= (q_{i+1}+\cdots\cdot+q_n)^2 \\
K_i &= \frac{s_{i-1,i}\; s_{i,i+1}}{s_{i-1,i+1}} \,.
\end{aligned}
\qquad (8.10)
$$

It is convenient to select as our independent variables the $s_{i,i+1}$, the t_i and K_i. There are, clearly, n-1 subenergies, n-1 momentum transfers, and n-2 K's: 3n-4 in all.

We expect that when all $s_{i,i+1}$ are large and all t_i are small, the amplitude takes the form[31)]

$$T = \beta_{AC_1}(t_1)\left(\frac{s_{12}}{s_o}\right)^{\alpha_1(t_1)} R_{C_2}(t_1,K_2,t_2)\cdot$$
$$\cdot\left(\frac{s_{23}}{s_o}\right)^{\alpha_2(t_2)} R_{C_3}(t_2,K_3,T_3)\cdots R_{C_{n-1}}(t_{n-2},K_{n-1},t_{n-1})\cdot$$
$$\cdot\left(\frac{s_{n-1,n}}{s_o}\right)^{\alpha_{n-1}(t_{n-1})} \beta_{BC_n}(t_{n-1})\cdot \qquad (8.11)$$

This is known as the MUREX (multi-Regge exchange) formula. The notation is as before. The trajectory $\alpha_i(t_i)$ is assumed to be the leading one having the quantum numbers of the channel $A\overline{C}_1\overline{C}_2\ldots\overline{C}_i$.

The kinematic region in which all subenergies are large and all momentum transfers are small is again seen to be the strong ordering region, where the rapidities of the produced particles satisfy $y_1 >> y_2 .. >> y_n$. Clearly, then $\omega_1 \approx \omega_n \approx \sqrt{s}/2$.

It should be clearly kept in mind that the particles $C_1 \ldots C_n$ which are produced do not themselves have to be stable; we could expect (8.11) to apply to the production of resonances, or clusters C_i as well as particles. All that is really required for (8.11) to apply is that the entities C have low masses.

The type of exclusive event described by (8.11) is illustrated on a rapidity plot in fig. 8.10. The event shows the production of 6 clusters, in which 4 clusters are single particles, one is a 2-particle cluster and one a

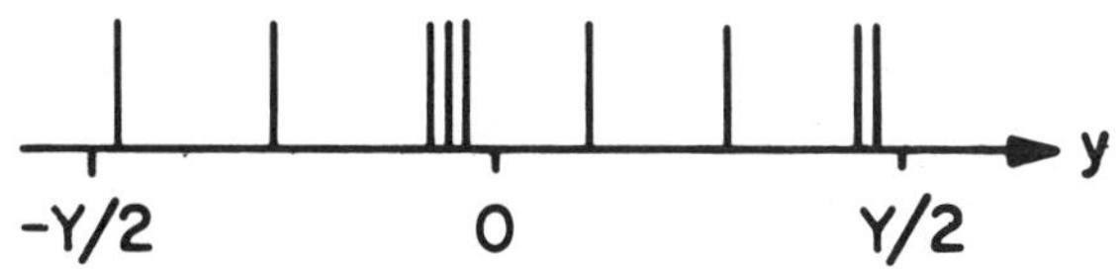

Fig. 8.10

3-particle cluster. The rapidity gaps between clusters is large; those within clusters are small. This event would be a multi-Regge event provided all momentum transfers between clusters are small, and would be described by (8.11) with n=6. (Not, we emphasize, with n=9.) The diagram as drawn is quite idealized. In practice the rapidity gap between particles coming from the decay of the same resonance is just about as large as that at which Regge exchange becomes reasonable. This, of course, is essentially only the statement that duality works. Therefore, the separation into clusters is, in practice, less clear than indicated in fig. 8.10.

We note that in the strong ordering region

$$s_{i,i+1} = m_{T_i} m_{T_{i+1}} e^{|y_i - y_{i+1}|}. \qquad (8.12)$$

Therefore, we have

$$\prod_{i=1}^{n-1} s_{i,i+1} = m_{T_1} m_{T_n} \prod_{i=2}^{n-1} m_{T_i}{}^2 e^{y_1 + y_n}$$

$$= \prod_{i=2}^{n-1} m_{T_i}{}^2 \, s \qquad (8.13)$$

and furthermore

$$K_i = \frac{s_{i-1,i}\ s_{i,i+1}}{s_{i-1,i+1}} = m_{T_i}^{\ 2}. \tag{8.14}$$

Because of (8.12), we may expect that for an n-particle MUREX reaction, each subenergy is of order $s^{1/(n-1)}$. If each of these is to be large, larger than some minimum value s_{min} at which we believe Reggeism sets in, then the necessary total energy is $s \sim (s_{min})^{n-1}$. Note, therefore, that $n = 1 + \ln s/s_{min}$, and hence the MUREX region contributes a term proportional to $\ln s$ to the average multiplicity, provided that it also contributes a finite fraction of the total cross section. Since $s \sim (s_{min})^{n-1}$, a very large amount of energy indeed is required for a high multiplicity event to be in the MUREX region. For example, if we believe $s_{min} \sim 4$ GeV^2, and we have $n=6$, we must have at least $s \sim 4^5 \sim 10^3$ GeV^2. Furthermore, if the fit we described before to $\pi^- p \rightarrow \pi^+ \pi^- \pi^- p$ at 25 GeV is any guide, only a small percentage of all six prong events at 10^3 GeV^2 will lie in the MUREX region. It is, consequently, not hard to see why essentially no careful fitting of data with the MUREX formula for n large has been done. We do not, therefore, need to inflict on the reader a detailed description of any horrendous pieces of phenomenology using MUREX.

The MUREX formula dredges up again our old difficulties with exact factorization and a

Pomeron having unit intercept. The contribution of the MUREX region of phase space alone provides a lower limit to the total cross section. If repeated Pomeron exchange can occur, and if $\alpha(o)=1$ for the Pomeron, then we can show[37] that this lower limit violates the Froissart bound, unless the internal Pomeron-Pomeron-particle vertex $R_C(t,K,t')$ vanishes as either t or t' goes to zero. It is, therefore, interesting to recall that in the Lipes et al. double-Regge fit we described earlier, there was no evidence for the existence of a Pomeron-Pomeron-f_0 vertex. However, the result can be strengthened to the statement that each internal Pomeron-Reggeon-particle vertex must also vanish as the t of the Pomeron goes to zero. This conclusion follows because a MUREX formula with alternating Pomerons and Reggeons also yields a lower limit to the total cross-section violating the Froissart bound. (We note that this coincides with what we found in chapter 6.) It is also worth noting that Lipes et al. did find evidence for the existence of a Pomeron-ρ Reggeon-ρ particle vertex in their analysis.

REFERENCES

1. See, e.g.,
P.D.B. Collins and E.J. Squires, "Regge Poles in Particle Physics," (Springer, Berlin 1968).
V. Barger and D. Cline, "Phenomenological Theories of High Energy Scattering," (Benjamin, New York 1969).
G.E. Hite, Rev. Mod. Phys. 41, 669 (1969).
J.D. Jackson, Rev. Mod. Phys. 42, 12 (1970).

P.D.B. Collins, Physics Reports 1C, 103 (1971).
2. M. Gell-Mann and B.M. Udgaonkar, Phys. Rev. Letters 8, 346 (1962).
3. P.G.O. Freund, Phys. Rev. Letters 20, 235 (1968).
H. Harari, Phys. Rev. Letters 20, 1395, (1968).
4. S.P. Denisov et al., Phys. Letters 36B, 415 (1971).
5. For example, see V. Barger, R.J.N. Phillips and K. Geer, Nucl. Phys. B47, 29 (1972).
6. V.N. Gribov, Nucl. Phys. 22, 249 (1961).
7. R. Oehme, Phys. Rev. Letters 18, 1222 (1967).
8. C.I. Tan and J. Finkelstein, Phys. Rev. Letters 19, 1061 (1967).
9. P.G.O. Freund and R. Oehme, Phys. Rev. Letters 10, 450 (1963).
10. J.S. Ball, G. Marchesini and F. Zachariasen, Phys. Letters 31B, 583 (1970).
11. A.H. Mueller, Phys. Rev. D2, 2963 (1970).
12 H.D.I. Abarbanel, Phys. Rev. D3, 2227 (1971).
H.M. Chan et al., Phys. Rev. Letters 26, 672 (1971).
13. M. Jacob, Rapporteur talk, Chicago Conf. 1972, Vol. 3, p. 373.
14. R. Brower, R. Cahn and J. Ellis, SLAC-PUB-1151, Nov. 1972.
15. T. Ferbel, Phys. Rev. Letters 29, 448 (1972).
16. M. Bishari, D. Horn and S. Nussinov, Nucl. Phys. B36, 109 (1972).
17. R. Peccei and A. Pignotti, Phys. Rev. Letters 27, 1538 (1971).
C.E. DeTar et al., Phys. Rev. D4, 906 (1971).
L. Caneschi and A. Pignotti, Phys. Rev. Letters 22, 1219 (1969).
R.P. Feynman in "High Energy Collisons," Eds. C.N. Yang et al. (Gordon and Breach, New York 1969) p. 237.
18. H.D.I. Abarbanel et al., Phys. Rev. Letters 26, 937 (1971).
19. M. Einhorn, J. Ellis and J. Finkelstein, Phys. Rev. D5, 2063 (1972).
20. C.E. Jones et al., Phys. Rev. D6, 1033 (1972).
21. D. Horn, Physics Reports 4C, no. 1, (1972).
22. C.E. DeTar et al., Phys. Rev. D4, 906 (1971).
23. M.G. Albrow et al., CHLM collaboration, Nucl. Phys., to be published.

24. F. Sannes et al., Phys. Rev. Letters 30, 766 (1973).
25. A. Capella, H. Högaasen and V. Rittenberg, SLAC-PUB-1176 (1973).
26. M.G. Albrow et al., CHLM collaboration proceedings of the Vanderbilt Conference 1973.
27. M.S. Chen, L.L. Wang and T.F. Wong, Phys. Rev. D5, 1667 (1972).
28. J. Ellis, J. Finkelstein and R. Peccei, Nuovo Cim. 12A, 763 (1972).
29. R. Brower and J. Weis, Phys. Letters 41B, 631 (1972).
30. H.D.I. Abarbanel et al., Phys. Rev. Letters 26, 937 (1971). Phys. Rev. D4, 2988 (1971).
31. T.W.B. Kibble, Phys. Rev. 131, 2282 (1963).
K.A. Ter Martirosyan, Zh. Ek. Teot. Fiz. 44, 341 (1963), Soviet Physics JETP 17, 233 (1963).
Chan Hong Mo, K. Kajantie and G. Ranft, Nuovo Cim. 49, 157 (1967).
F. Zachariasen and G. Zweig, Phys. Rev. 160, 1322, 1326 (1967).
32. C.E. DeTar and J.H. Weis, Phys. Rev. D4, 3141 (1971).
33. O. Steinmann, Helv. Phys. Acta 33, 257 (1960); 33, 349 (1960).
34. D. Silverman and C.I. Tan, Phys. Rev. D1, 3479 (1970).
35. M. Aderholz et al., ABBBHLM collaboration, Nuovo Cim. 34, 495 (1964).
36. R. Lipes, G. Zweig and W. Robertson, Phys. Rev. Letters 22, 433 (1969).
37. J. Finkelstein and K. Kajantie, Physics Letters 26B, 305 (1968).

PART III

FIELD THEORETICAL MODELS

9

Ladders in ϕ^3 Field Theory: Regge Poles

Field theory provides, at present, the only well defined theory we have from which we can, in principle at least, calculate cross sections theoretically at very high energies. Unfortunately, we do not know which field theory to use. Surely we are not going to believe in a field theory containing each one of the panoply of observed hadrons as an elementary constituent; indeed, such a theory is not even renormalizable. The only plausible idea is to think of the hadrons as composed of some basic constituents, such as quarks, and to write a field theory for these. Still, we do not know what field theory to write down. The quark-vector gluon model is perhaps the obvious guess, but few people are likely to take its detailed predictions seriously. It may suffice to imitate the real world near the light cone, but there is no reason that it should do so for very high energy cross sections.

The only field theory in which we actually have confidence is quantum electrodynamics, and even here there is no experimental confirmation at extreme energies. Furthermore, there is no obvious reason to believe that the behavior of a theory of photons and electrons coincides with the behavior of hadrons.

All of this means that we don't really have any reason to expect field theory to give us the right answers; nevertheless, since field theory is at least relatively well defined, it provides the natural starting point for our theoretical discussion of very high energy hadronic processes.

Having said the above, it would now be nice to proceed to derive the asymptotic properties of a field theory. Needless to say we cannot do this in practice. Not only do we not know that field theory is relevant; we can't calculate its predictions anyway. All that we'll be able to do is to sum certain classes of Feynman diagrams, and hope that their behavior approximates the true behavior of the theory.

Since we do not really know which field theory to use, we will start with the simplest one, containing a single boson of mass μ described by a field ϕ, with a coupling $g\phi^3$. There are theoretical problems with such a theory, for example that it has no lowest energy; we shall ignore these.

For elastic scattering, in lowest order there are three diagrams and the amplitude is

$$T(s,t) = \frac{g^2}{s-\mu^2} + \frac{g^2}{t-\mu^2} + \frac{g^2}{u-\mu^2} . \tag{9.1}$$

As $s \to \infty$ with t fixed, the exchange diagram dominates, and

$$T(s,t) \to \frac{g^2}{t-\mu^2} . \tag{9.2}$$

Therefore $\sigma_2 = \sigma_{e\ell} \propto s^{-2}$ in this order.

For the two-particle into many-particle reaction, there are many diagrams; in the kinematic region where s as well as all subenergies are large, but where momentum transfers are fixed, the dominant diagrams in lowest order are the peripheral ones. Let us guess, therefore, that for the n-particle production amplitudes, the dominant diagrams are multiperipheral:

$$T_{2\to n} \to g\frac{1}{t_1-\mu^2}\, g\frac{1}{t_2-\mu^2}\, g \cdots \frac{1}{t_{n-1}-\mu^2}\, g \tag{9.3}$$

when $t_1 .. t_{n-1}$ are finite and all subenergies go to infinity. The corresponding diagram appears in fig. 9.1.

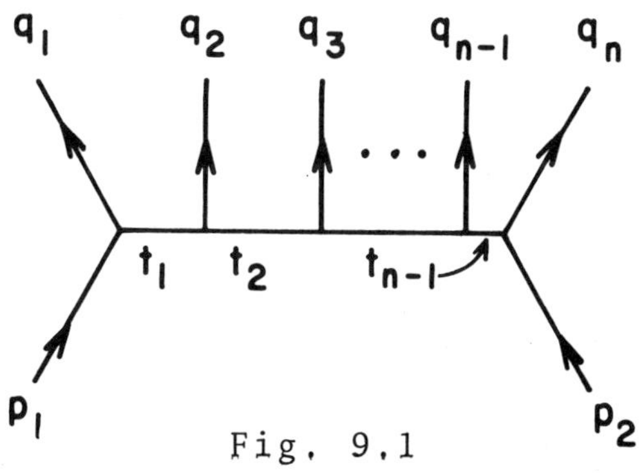

Fig. 9.1

Two qualitative conclusions are immediate, and these are not particularly encouraging. First, there is a cutoff on each t, and hence on q_T, provided by the Feynman propagators. But this cutoff is only proportional to 1/t, which is not nearly sharp enough to agree with experiment. Perhaps this will be fixed by higher orders in g. Second, the power dependence of the cross sections is s^{-2}, which is also clearly wrong. This is perhaps to be cured by using a field theory containing spin, as well as by adding higher orders in g.

If it is indeed true that the graph of fig. 9.1 is a plausible starting point for $T_{2\to n}$, then we may use this knowledge together with s-channel unitarity to calculate higher order corrections to the elastic amplitude. This suggests that we should look at the set of ladder graphs for a significant set of higher order corrections.

To study higher orders in g at large s, we evidently need an approximation procedure, since we cannot calculate the ladder exactly. The procedure usually invoked[1] is the "leading ℓn approximation," which consists of guessing that in a given order in g we need keep only the most divergent term in s. This is clearly a highly dangerous procedure: It would have us believe that we can approximate $a_2g^2+g^4(a_4\ell n s+ b_4)+g^6(a_6\ell n^2 s+b_6\ell n s+c_6)+\ldots$ by $a_2g^2+a_4g^4\ell n s + a_6g^6\ell n^2 s+\ldots.$ neglecting $b_6g^2 \ell n s$ compared to $a_4g^4 \ell n s$ and a_2g^2.

Let us illustrate the use of this procedure by applying it to the calculation of the ladder.

In lowest order, the ladder gives

$$T_1 = \frac{g^2}{s-\mu^2} \to \frac{g^2}{s} . \tag{9.4}$$

In fourth order, we have the box (fig. 9.2a), which has the behavior $g^4 \ln s/s$.[2] In $2n^{th}$ order (fig. 9.2b), the ladder graph turns out[3] to be proportional to $g^{2n}(\ln s)^{n-1}/s$.

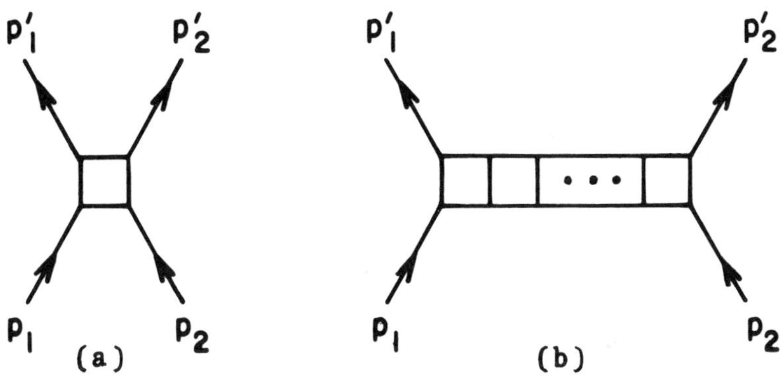

Fig. 9.2

If we add all this up, we have the form

$$T_{\text{ladder}}(s,t) \to \sum_{n=1}^{\infty} g^{2n} \, g_n(t) \, \frac{(\ln s)^{n-1}}{s}. \tag{9.5}$$

The calculation of the coefficients $g_n(t)$ is relatively straightforward and can be carried out in several different ways. Let us begin by describing a technique using dispersion theory;[4] we shall outline the direct calculation of the set of Feynman diagrams later.

The dispersion approach approximates the ladder by retaining only the two-particle intermediate state contributions in the t-channel. We first note that in this approximation the Mandelstam double spectral function[5] for the nth order ladder can be written

$$\rho_n(s,t) = \sum_{m=1}^{n-1} \frac{1}{8\pi^2} \frac{1}{\sqrt{t(t-4\mu^2)}} \iint \frac{ds_1\, ds_2}{\sqrt{\lambda(s,s_1,s_2,t)}} \cdot$$

$$\cdot\; A_m^*(s_1,t)\, A_{n-m}(s_2,t) \tag{9.6}$$

where A_m is the s-channel absorptive part of the mth order ladder. Thus

$$A_m(s,t) = \frac{1}{\pi}\int \frac{dt'}{t'-t}\, \rho_m(s,t). \tag{9.7}$$

The function λ is the usual kinematic kernel: We have

$$\lambda = s^2+s_1{}^2+s_2{}^2-2ss_1-2ss_2-2s_1s_2-\frac{4ss_1s_2}{t-4\mu^2}. \tag{9.8}$$

The relation is illustrated graphically in fig. 9.3.

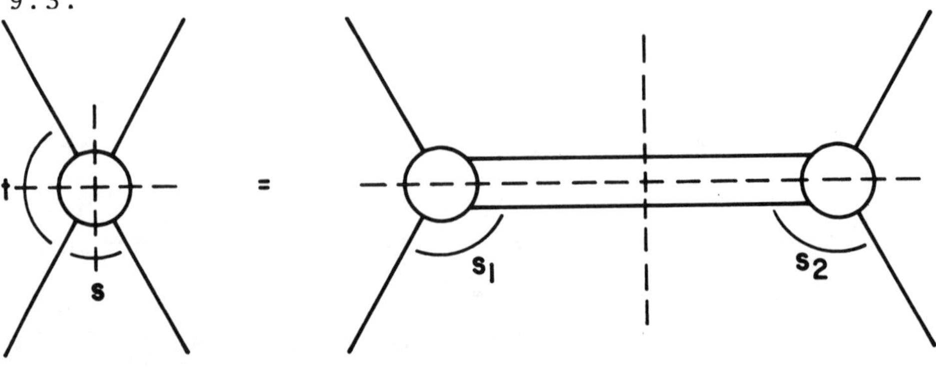

Fig. 9.3

Eq. (9.6) can be used to generate the entire perturbation series for the ladder by induction.[4] The starting point is the double spectral function for the lowest order box, which is easily found to be asymptotically

$$\rho_2(s,t) = \frac{g^4}{8s} \frac{1}{\sqrt{t(t-4\mu^2)}} \Theta(t-4\mu^2). \tag{9.9}$$

To find the succeeding orders let us guess the form of the final result; namely that the entire amplitude can be written

$$T_{ladder}(s,t) = \frac{g^2}{s} (-s)^{g^2 f(t)}. \tag{9.10}$$

The mth order term is then

$$T_m(s,t) = \frac{g^2}{s} \frac{(g^2 f(t) \ell n(-s))^{m-1}}{(m-1)!} \tag{9.11}$$

and the mth order s-channel absorptive part is, in the leading ℓn approximation,

$$A_m(s,t) = \frac{g^2}{s} \pi g^2 f(t) \frac{(g^2 f(t) \ell n\ s)^{m-2}}{(m-2)!}. \tag{9.12}$$

The function f(t) is found by comparing with lowest order: We have from (9.12) that

$$A_2(s,t) = \frac{\pi g^4}{s} f(t) \tag{9.13}$$

so that

$$\rho_2(s,t) = \frac{\pi g^4}{s} \operatorname{Im} f(t); \tag{9.14}$$

to agree with (9.9) we must have

$$\operatorname{Im} f(t) = \frac{1}{8\pi} \frac{1}{\sqrt{t(t-4\mu^2)}} \Theta(t-4\mu^2). \tag{9.15}$$

If we now suppose these equations to be valid for $m \leq n-1$, we can use eq. (9.6) to calculate the nth order term, and to verify that it also has the anticipated form. If it does we will have confirmed our guess by induction. That the induction in fact works is not difficult to check. We insert (9.12) into (9.6), evaluate the integral for large s retaining only the leading powers of $\ell n\ s$, and find

$$\rho_n(s,t) = \frac{g^4}{8s} \frac{1}{\sqrt{t(t-4\mu^2)}} \frac{(\ell n\ s)^{n-2}}{(n-2)!} \cdot \frac{\operatorname{Im}(g^2 f(t))^{n-1}}{\operatorname{Im} f(t)} \tag{9.16}$$

which using (9.15), is precisely what we want. Eq. (9.10) is thereby justified.

The entire result, in the leading ℓn approximation, is simply[3,4]

$$T_{ladder}(s,t) \to \beta(t)(-s)^{\alpha(t)}$$

where

$$\alpha(t) = -1+g^2 f(t) = -1+\frac{g^2}{8\pi^2} \frac{1}{\sqrt{t(t-4\mu^2)}} \cdot$$

$$\cdot \ \ln \frac{\sqrt{t-4\mu^2} + \sqrt{t}}{\sqrt{t-4\mu^2} - \sqrt{t}} \tag{9.17}$$

and $\beta(t) = g^2$.

Note that as $t\to\pm\infty$, $\alpha(t)\to -1$ and that $\alpha(o) = -1+\frac{g^2}{16\mu^2\pi^2}$. In the leading $\ln$ approximation the ladder generates a Regge pole, with trajectory and residue calculated to order g^2; the trajectory, in the limit of vanishing g^2, lies at $j=-1$.

From (9.17) we see that the total cross section from the ladder is

$$\sigma_T(s) = \pi g^4 f(o) s^{-2+g^2 f(o)} \tag{9.18}$$

it can, therefore, be made to be a constant, for example, by appropriately adjusting g^2.

The total cross section is made up from partial cross sections for producing n particles which can be read off from eq. (9.12); we have

$$\sigma_n(s) = \pi g^4 f(o) s^{-2} \frac{(g^2 f(o) \ln s)^{n-2}}{(n-2)!}. \tag{9.19}$$

The average multiplicity is therefore

$$\langle n(s) \rangle = g^2 f(o) \ln s + 2. \tag{9.20}$$

All of these results are the implications of the model of the production amplitude given in eq. (9.3), and in fig. 9.1.

Precisely the same conclusions[6)] follow from calculating the entire Feynman diagram for the ladder in the leading ℓn approximation. This fact confirms the dispersion method result and tells us that the retention of only two-particle cuts in the t-channel correctly gave the asymptotic behavior in this approximation. A brief outline of the evaluation of the ladder directly from Feynman diagrams follows.[6)] The amplitude for the lowest order box of fig. 9.2a is

$$T_2(s,t) = g^4 \int \frac{d^4q}{(2\pi)^4} \frac{1}{(p_1-q)^2-\mu^2} \frac{1}{(p_1'-q)^2-\mu^2} \cdot$$

$$\cdot \frac{1}{q^2-\mu^2} \frac{1}{(P-q)^2-\mu^2} \tag{9.21}$$

where $P^2=(p_1+p_2)^2=s$ and $(p_1'-p_1)^2=t$. After introducing Feynman parameters and integrating over q, this may be rewritten

$$T_2(s,t) = \frac{g^4}{16\pi^2} \int_0^1 \frac{d\alpha_1 d\alpha_2 d\beta_1 d\beta_2 \delta(1-\alpha_1-\alpha_2-\beta_1-}{(\alpha_1\alpha_2 s+d(\alpha_1\alpha_2\beta_1\beta_2 t))^2} \tag{9.22}$$

where $d = \beta_1\beta_2 t+(\alpha_1\beta_1+\alpha_1\beta_2+\alpha_2\beta_1+\alpha_2\beta_2)\mu^2$

$$-(\alpha_1+\alpha_2+\beta_1+\beta_2)^2\mu^2. \tag{9.23}$$

Now the dominant contribution to the integral when s is large occurs if α_1 or α_2 is near zero.[6] Let us isolate the region where α_1 and α_2 are less than ε. There we can neglect α_1 and α_2 in d. The α_1 and α_2 integrals can then be carried out, and yield the result

$$T_2(s,t) = g^4 \, f(t) \, \frac{\ln(-s)}{s} \tag{9.24}$$

independent of ε, where

$$f(t) = \frac{1}{16\pi^2} \int_0^1 d\beta_1 \int_0^1 d\beta_2 \, \cdot \, \frac{\delta(\beta_1+\beta_2-1)}{\beta_1\beta_2 t - (\beta_1+\beta_2)^2 \mu^2} \, . \tag{9.25}$$

This expression can be shown to coincide with what we found earlier, so labeling it f(t) is indeed justified.

The general term in the ladder can be evaluated in much the same way. One finds that the n rung contribution is[6]

$$T_n(s,t) = g^2 \left(\frac{g^2}{16\pi}\right)^{n-1} (n-1)! \, \cdot \, \int \frac{\Pi d\alpha d\beta \;\; \delta(\Sigma\alpha+\Sigma\beta-1) \, [C(\alpha,\beta)]^{n-2}}{[\alpha_1 \ldots \alpha_n s + d(\alpha,\beta,t)]^n} \, ; \tag{9.26}$$

this is again dominated by the region where some α is small, and gives

$$T_n(s,t) \to g^2 \frac{1}{(n-1)!} (g^2 f(t)\ln(-s))^{n-1}. \tag{9.27}$$

We, therefore, recover our earlier results.

If we ask how to evaluate the set of ladder graphs exactly in the large s limit, without invoking the leading $\ln$ approximation, we are faced with the question of solving the Bethe-Salpeter equation[7)] with a single rung as input. Evidently, the off mass-shell ladder amplitude satisfies

$$T_{\text{ladder}}(p_1 p_2 \to p_1' p_2') = \frac{g^2}{s-\mu^2} + \int \frac{d^4q}{(2\pi)^4} \frac{g^2}{q^2-\mu^2} \cdot \frac{1}{(p_1-q)^2-\mu^2} \frac{1}{(p_1'-q)^2-\mu^2} \cdot T_{\text{ladder}}(p_1-q,\ p_2 \to p_1'-q,\ p_2') \tag{9.28}$$

The on shell ladder amplitude is simply $T_{\text{ladder}}(p_1 p_2 \to p_1' p_2')$ at the point $p_1^2 = p_2^2 = p_1'^2 = p_2'^2 = \mu^2$.

Analytic solutions of this equation are unknown, though it is amenable to numerical techniques: We shall come back to some numerical consequences of it shortly.

While the Bethe-Salpeter equation cannot be solved exactly it is, nevertheless, possible to prove something about the solution. It can be demonstrated that as $s \to \infty$ the solution is a Regge pole, and it can be demonstrated that as $g^2 \to 0$ this pole approaches -1. This again tends to confirm the at least qualitative relevance of the leading ℓn approximation.

As an illustration of the results obtained from numerically solving the Bethe-Salpeter equation, we display fig. 9.4. This shows what happens not in the ϕ^3 theory, but in a theory with π exchange and ρ rungs in the ladder.[8)] The $\pi\pi\rho$

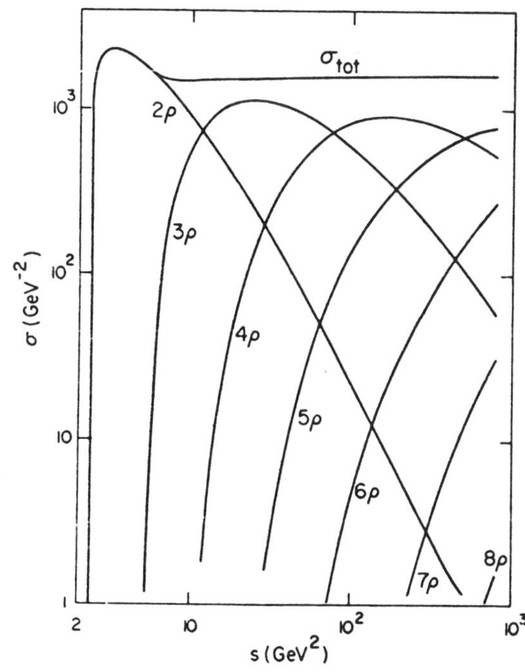

Fig. 9.4. Results of a multiperipheral calculation by Wyld (reference 8) of the process $\pi\pi \to n\rho$, $n=2,3,\ldots$, based on elementary pion exchange. The only free parameter was the $\rho\pi\pi$ coupling which was chosen so as to give an asymptotic constant total cross section.

coupling is adjusted so that the Regge trajectory generated by the ladder passes through 1 at t=0 to give a constant total cross section. The way in which this is built up from σ_n is shown; everything is much like what we outlined in eqs. (9.18) and (9.19), for the ϕ^3 case. The value of the coupling needed to produce a constant cross section also predicts its value, a la eq. (9.18): 587 mb results.[8)] Evidently, these field theories are not to be taken literally.

10

Iterated Ladders in ϕ^3 Field Theory: Regge Cuts

What lies beyond the ladder? We at once think of two ladders, like those shown in fig. 10.1.

A first attempt at estimating the high energy behavior of this set of graphs is to retain only the two-body contribution to the s-channel absorptive part.[9] This approximation yields

$$A(s,t) = \frac{1}{16\pi^2 s}\iint \frac{dt_1\, dt_2}{\sqrt{-\lambda(t,t_1,t_2)}} T_{\text{ladder}}(s,t_1)\cdot T^*_{\text{ladder}}(s,t_2) \quad (10.1)$$

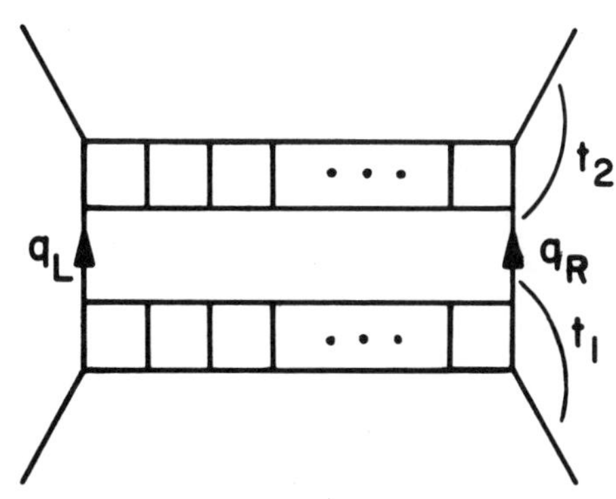

Fig. 10.1

for the absorptive part at large s.

If $T_{\text{ladder}}(s,t)$ is indeed a Regge pole, say of even signature, then eq. (10.1) generates a Regge cut. For, suppose we take

$$T_{ladder}(s,t) = \gamma(t)s^{\alpha(t)}e^{-i\pi\alpha(t)/2} \tag{10.2}$$

and for simplicity let $\gamma(t)=\gamma_0 e^{b_0 t}$ and $\alpha(t)=\alpha_0+\alpha' t$. Then we find from (10.1) that as $s\to\infty$

$$A(s,t) \to \frac{|\gamma_0|^2}{32\pi\alpha'} e^{\frac{b_0}{2}t} \frac{s^{\alpha_c(t)}}{\ell n\ s} \tag{10.3}$$

$$\text{where } \alpha_c(t) = 2\alpha(\tfrac{t}{4}) - 1. \tag{10.4}$$

This corresponds to a t-channel j-plane amplitude containing a logarithmic branch point at $j=\alpha_c(t)$ (see appendix D).

This, it is easy to see, is true even if we do not use the above simple forms for $\gamma(t)$ and for $\alpha(t)$. To demonstrate this, let us project eq. (10.1) into t-channel partial waves through the usual definitions

$$T(s,t) = \frac{1}{2\pi i}\int_{-i\infty}^{i\infty} dj\ s^j e^{-i\pi j/2}\ \cdot$$

$$\cdot \left\{-\frac{T^+(t,j)}{\sin\frac{\pi j}{2}} + i\frac{T^-(t,j)}{\cos\frac{\pi j}{2}}\right\} \tag{10.5}$$

$$\text{and}\quad A(s,t) = \frac{1}{2\pi i}\int_{-i\infty}^{i\infty} dj\ s^j \left\{T^+(t,j)+T^-(t,j)\right\}. \tag{10.6}$$

Taking $T^-_{ladder}(t,j)=0$ and $T^+_{ladder}(t,j)=\beta(t)/(j-\alpha(t))$, we find that the double ladder of fig. 10.1 gives

$$T^+(t,j)+T^-(t,j) = \frac{1}{16\pi^2}\iint \frac{dt_1 dt_2}{\sqrt{-\lambda}} \cdot$$

$$\cdot \frac{\gamma(t_1)\ \gamma(t_2)}{j-\alpha(t_1)-\alpha(t_2)+1}\ e^{-\frac{i\pi}{2}(\alpha(t_1)-\alpha(t_2))} \quad (10.7)$$

where we define $\gamma(t)=\beta(t)/(\sin \pi\alpha(t)/2)$. Evidently, the branch point exists at $\alpha_c(t)=2\alpha(t/4)-1$ provided $\alpha(t)$ is monotonically increasing. Note also, however, that it occurs in both signatured partial wave amplitudes; the cut has both signatures. This is a serious failure, for if $T(t,j)$ has a cut in j which varies with t it also has a cut in t which varies with j. But it, therefore, produces a branch point in t even in a physical partial wave amplitude, and this branch point does not, in general, coincide with a branch point dictated by unitarity. Only if the cut has a given signature is this difficulty resolved,[10] for then the discontinuity across the cut vanishes at sense integers, so that in physical partial wave amplitudes the cut is absent. Our concern, in studying diffraction, is with even signatured amplitudes. Hence an odd signatured cut must be rejected.

This cut is known as the AFS cut.[9] In its derivation we retained only the two-particle intermediate states and this fact was crucial in obtaining the cut. If one calculates the Feynman diagrams of fig. 10.1 completely they involve, of course, many particle intermediate states as well as the two-particle state. It turns out that all these intermediate states conspire to cancel out the AFS cut.[10)] This follows because one can show on general grounds that no planar Feynman diagram has the asymptotic AFS cut form. However, if one considers nonplanar diagrams, as for instance the set of diagrams shown in fig. 10.2, then a cut reappears - it is now known as the Mandelstam cut.

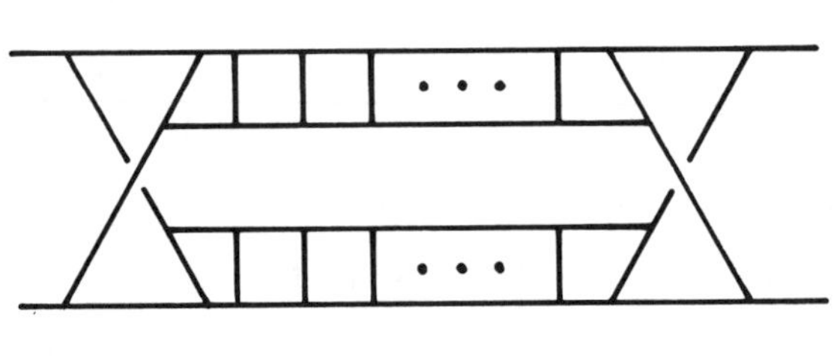

Fig. 10.2

It is worth understanding why this cancellation of the AFS cut occurs in the diagram of fig. 10.1 while it does not in the diagram of fig. 10.2.[11)]

When fig. 10.1 is regarded as a Feynman diagram, it is necessary to integrate over all four momenta q_L and q_R of the left- and right-hand center lines. The diagram will have the form

$$\int \frac{dq_L{}^2}{q_L{}^2-\mu^2+i\varepsilon} \int \frac{dq_R{}^2}{q_R{}^2-\mu^2+i\varepsilon} \iint \frac{dt_1 dt_2}{\sqrt{-\lambda}} \cdot$$

$$\cdot\, T_{\text{ladder}}(s,t_1,q_L{}^2,q_R{}^2) \cdot$$

$$\cdot\, T^*_{\text{ladder}}(s,t_2,q_L{}^2,q_R{}^2). \qquad (10.8)$$

The two-body intermediate state contribution, which led to eq. (10.1), is obtained by replacing the propagator by delta functions here:

$$\frac{1}{q_L^2-\mu^2+i\varepsilon} \to -i\pi\delta(q_L^2-\mu^2) \text{ and similarly for } q_R.$$

However, when the integral over dq_L^2 and dq_R^2 is kept, then one sees that the entire diagram vanishes. This is because the integral over dq_L^2, for example, runs from $-\infty$ to $+\infty$, contains a pole at $q_L^2=\mu^2-i\varepsilon$ and a right-hand cut only which the contour of integration passes above. Therefore, the contour of integration can be closed in the upper half q_L^2 plane, (assuming that T_{ladder} vanishes at large q_L^2); since no singularities are enclosed by the contour, the integral vanishes. In the diagram of fig. 10.2, in contrast, there is both a right-hand and a left-hand cut in q_L^2, and the contour of integration lies above the right-hand cut but below the left-hand cut. Hence the contour cannot be closed in the upper half plane alone, and the integral will not vanish.

The crucial ingredient in having a nonvanishing contribution is the existence of a left-hand cut.[10] That requires diagrams, such as fig. 10.2 for example, which are nonplanar. Nonplanar diagrams like fig. 10.2 can be divided in two to yield diagrams for production amplitudes like those shown in fig. 10.3(a). Such diagrams are called polyperipheral,[12] and should be contrasted with the multiperipheral diagrams shown in fig. 10.3(b), which are obtained by breaking a single ladder graph in half. The production mechanism associated with ladders is multiperipheral; that

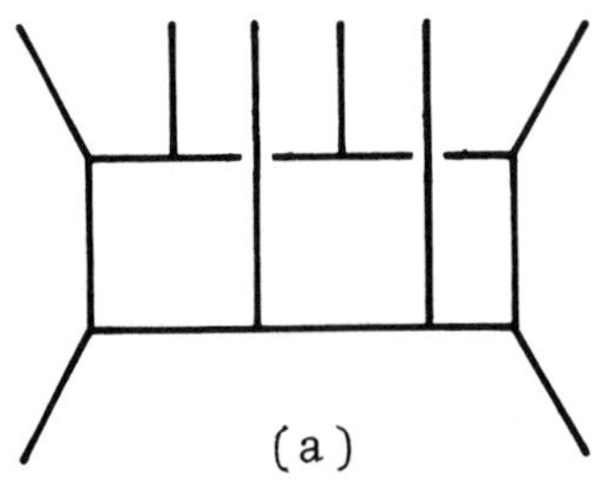

(a)

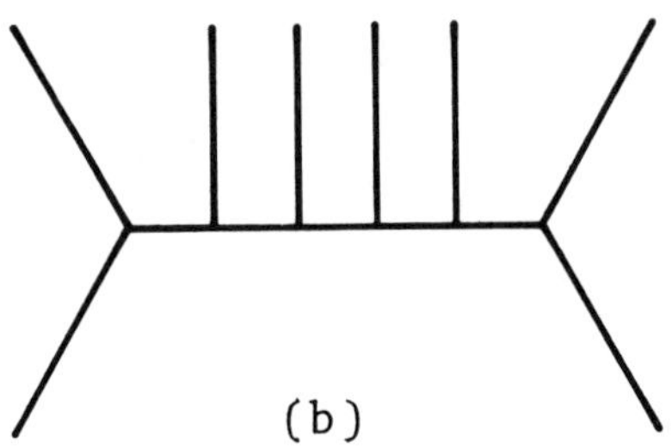

(b)

Fig. 10.3

associated with the Mandelstam cut is polyperipheral.

Rather than attempt to derive the behavior of the diagrams of fig. 10.2 directly, we shall first discuss the repeated exchange of many ladders. If we understand what n-ladder exchange gives, then we can as a special case find out what to expect from two-ladder exchange.

The repeated exchange in the s-channel of any structure leads us into the subject of eikonals, and we shall begin our discussion of this subject with the simplest example, namely that in which the exchanged structure is just a single particle.[13] Let us, therefore, turn to the set of diagrams shown in fig. 10.4. That is, we want to discuss an s-channel ladder, in which the rungs can be crossed at will.

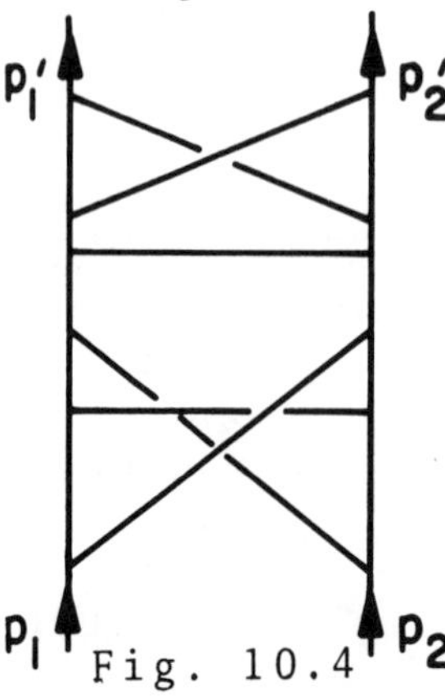

Fig. 10.4

The scattering amplitude for the diagram with n rungs is, according to the usual Feynman rules,

$$T_n(s,t) = g^{2n}\prod_{i=1}^{n}\int\frac{d^4k_i}{(2\pi)^4}\,\frac{1}{k_i^{\,2}-\mu^2}\,(2\pi)^4\,\cdot$$

$$\cdot\,\delta^4(p_1-p_1'-\Sigma k)\left\{\frac{1}{(p_1-k_1)^2-\mu^2}\,\cdot\right.$$

$$\left.\cdot\,\frac{1}{(p_1-k_1-k_2)^2-\mu^2}\cdots\frac{1}{(p_1-k_1-k_2-\ldots k_n)^2-\mu^2}\right\}\cdot$$

$$\cdot\sum_{\text{perm. of }1\ldots n}\left\{\frac{1}{(p_2+k_1)^2-\mu^2}\cdots\cdot\right.$$

$$\left.\cdot\,\frac{1}{(p_2+k_1+\ldots.+k_n)^2-\mu^2}\right\} \tag{10.9}$$

where $k_1..k_n$ are the momenta of the exchanged particles in the order in which they are emitted from particle number 1.

We may use the delta function to eliminate one k; to preserve the symmetry of the expression we add the expression n times, once with each k eliminated, and divide by n. In a given such term, with k_r the eliminated momentum, let s be the position on particle 2 where k_r is absorbed. Let $\bar{k}_1..\bar{k}_{s_1}$, be the set of $k_1..k_{r-1}$ which are absorbed on particle 2 before s, and $\bar{k}_{s_1+1}..\bar{k}_{s_1+s_2}$ be the set of $k_{r+1}...k_n$ absorbed before s, with $s_1+s_2=s-1$. Then the sum over permutations in eq. (10.9) is: the sum over all sets $\{s_1\}$ and $\{s_2\}$, and the sum over all permutations P_2 of the

$\bar{k}_1 \ldots \bar{k}_{s_1+s_2}$ and permutations P_2' of the remining $\bar{k}$'s. For the sake of symmetry, one can also sum over all permutations P_1 of $k_1 .. k_{r-1}$ and p_1' of $k_{r+1} .. k_n$ and divide by $(r-1)!\,(n-r)!$; each of these is just a relabeling of the k's.

Now let us introduce the high energy assumption. This states that for large s and near the forward direction, a propagator like $(p_1 \pm k)^2 - m^2$ can be replaced by $\pm 2p_1 \cdot k$. That is, we neglect k^2 compared to $2p \cdot k$ in the very high energy very small angle limit. This is the crux of the eikonal approximation. It means essentially that the particles numbers 1 and 2 do not suffer any appreciable recoil while absorbing and emitting the various exchanged particles. This will be true if the dynamics of the system calls for very small exchanged momenta. Such a situation arises in the study of Bremsstrahlung in QED where the sum of all radiation diagrams describes a classical wave emission from a heavy point source. The classical limit for scattering corresponds to the approximation of using geometrical optics, which is where the eikonal method acquires its name. In the present situation, however, it may be only a crude approximation.

In this spirit let us rewrite eq. (10.9) in the form

$$T_n(s,t) = g^{2n}\,\frac{1}{n}\sum_{r=1}^{n}\int\frac{d^4k_1}{(2\pi)^4}\cdots\int\frac{d^4k_{r-1}}{(2\pi)^4}\,\cdot$$

$$\cdot\int\frac{d^4k_{r+1}}{(2\pi)^4}\cdots\int\frac{d^4k_n}{(2\pi)^4}\prod_{i=1}^{n}\frac{1}{k_i^2-\mu^2}\,\cdot$$

$$\cdot\ \frac{1}{(r-1)!\,(n-r)!}\sum_{\text{sets }\{s_1\}\{s_2\}}\left\{\sum_{P_1P_1'}\frac{1}{a_1}\,\cdot\right.$$

$$\cdot\ \frac{1}{a_1+a_2}\cdots\frac{1}{a_1+\ldots+a_{r-1}}\,\frac{1}{a_n'}\cdots\frac{1}{a_n'+\ldots+a_{r+1}'}\,\cdot$$

$$\cdot\ \sum_{P_2P_2'}\frac{1}{b_1}\,\frac{1}{b_1+b_2}\cdots\frac{1}{b_1+\ldots+b_{s-1}}\,\frac{1}{b_n'}\cdots\,\cdot$$

$$\left.\cdot\ \frac{1}{b_n'+\ldots+b_{s+1}'}\right\} \tag{10.10}$$

where $a_i = -2p_1\cdot k_i \qquad b_i = 2p_2\cdot\bar{k}_i$

$$a_i' = 2p_1'\cdot k_i \qquad b_i' = 2p_2'\cdot\bar{k}_i\,. \tag{10.11}$$

The sums over permutations can be carried out and allow us to replace the brackets in eq. (10.10) by

$$\frac{1}{a_1\ldots a_{r-1}}\,\frac{1}{a_{r+1}'\ldots a_n'}\,\frac{1}{b_1\ldots b_{s-1}}\,\frac{1}{b_{s+1}'\ldots b_n'}.$$

Next rewrite the rth propagator in eq. (10.10) as follows

$$\frac{1}{k_r^2-\mu^2} = \int d^4x \; \Delta_F(x) \; e^{-i(p_1'-p_1)\cdot x} \; .$$

$$\cdot \; e^{i(k_1+..+k_{r-1})\cdot x} \; e^{i(k_{r+1}+..+k_n)\cdot x}. \tag{10.12}$$

Thus $$T_n(s,t) = \frac{1}{n} \sum_r \frac{1}{(r-1)!} \frac{1}{(n-r)!} \sum_{\mathrm{sets}\{s_1\}\{s_2\}} \cdot$$

$$\cdot \int d^4x \quad \Delta_F(x) \; e^{-i(p_1'-p_1)\cdot x} \; .$$

$$\cdot \quad U_{++}^{s_1} \; U_{+-}^{r-1-s_1} \; U_{-+}^{s_2} \; U_{--}^{n-r-s_2} \tag{10.13}$$

where $$U_{\pm\pm} = U\begin{pmatrix} x; & +p_1 & +p_2 \\ & -p_1' & -p_2' \end{pmatrix} \tag{10.14}$$

and where

$$U_{++}(x,p_1,p_2) = \prod_{\substack{i=1 \\ i\neq r}}^{n} g^2 \int \frac{d^4k_i}{(2\pi)^4} \frac{1}{k_i^2-\mu^2} \frac{1}{2p_1\cdot k} \cdot$$

$$\cdot \; \frac{1}{2p_2\cdot k} \quad e^{ik_ix}. \tag{10.15}$$

Now the number of sets $\{s_1\}$ and $\{s_2\}$ is $(r-1)!/[s_1!(r-1-s_1)!]$ and $(n-r)!/[s_2!(n-r-s_2)!]$ and the contribution of each set is identical. All the sums may, therefore, be performed in eq. (10.13) to yield

$$T(s,t) = \int d^4x \; e^{-i(p_1'-p_1)\cdot x} g^2 \Delta_F(x) \left(\frac{e^{2i\chi}-1}{2i\chi} \right) \tag{10.16}$$

where the function χ, which is known as the eikonal, is given by

$$2i\chi = -i(U_{++}+U_{+-}+U_{-+}+U_{--}). \tag{10.17}$$

From (10.17) we see that χ depends only on $p_1 \cdot p_2$ - namely s - and $p_1 \cdot x$ and $p_2 \cdot x$. In the high energy forward regime, the integral over d^4x in (10.16) can be reduced to a single integral over the component of x_μ transverse to the direction of motion. This transverse distance is called the impact parameter and is denoted b. Eq. (10.16), then, can be rewritten in the form[13)]

$$T(s,t) = 8\pi s \int_0^\infty bdb \; J_o(b\sqrt{-t}) \left(\frac{e^{2i\chi(s,b)}-1}{2i} \right). \tag{10.18}$$

We may calculate χ explicitly from (10.17), but it is easier to obtain it from (10.18) by expanding both sides of the equation in powers of g^2. To lowest order, the left-hand side of (10.18) is simply one-particle exchange, so that we find

$$\frac{g^2}{t-\mu^2} = 8\pi s \int bdb\ J_0(b\sqrt{-t})\ \chi(s,b); \qquad (10.19)$$

this determines χ.

Eq. (10.18) by itself, of course, merely expresses the scattering amplitude in terms of its Bessel transform; such a representation can obviously be written for any amplitude. The content of the particular set of graphs of fig. 10.4 lies in eq. (10.19), which gives the eikonal χ for these graphs. With this choice of χ, the eikonal formula (10.19) is completely analogous to the well known nonrelativistic Glauber formula[14)] with $g^2/(t-\mu^2)$ playing the role of a potential.

More generally, for an arbitrary eikonal χ, the phase shift δ and the absorption η are defined by writing

$$\chi(s,b) = \delta(s,b) - \frac{i}{2}\ \ell n\ \eta(s,b) \qquad (10.20)$$

$$\text{thus}\quad T(s,t) = 8\pi s \int bdb\ J_0(b\sqrt{-t})\ \frac{\eta e^{2i\delta} - 1}{2i} \qquad (10.21)$$

as usual.

The next question to be asked is: Can we write a formula like (10.18) for the exchange of a structure other than a single particle? For example, does eq. (10.18), but with χ given by

$$T_{ladder}(s,t) = 8\pi s \int bdb\ J_0(b\sqrt{-t})\ \chi(s,b) \qquad (10.22)$$

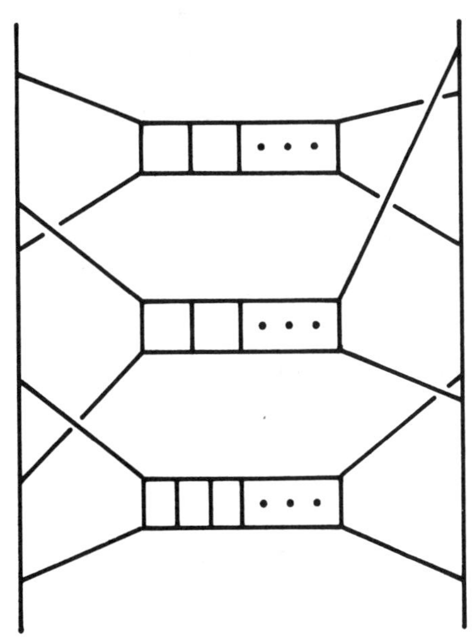

Fig. 10.5

correspond to the set of diagrams shown in fig.10.5? The answer is, in general, yes.[15)]

Let us first expand eq. (10.18) in powers of χ. We find

$$T(s,t) = T_{ladder}(s,t) + 8\pi is \int_0^\infty bdb\ J_o(b\sqrt{-t}) \cdot$$

$$\cdot\ (\chi(s,b))^2 + \ldots \qquad (10.23)$$

and using (10.22) together with eq. (E7) from appendix E, this yields

$$T(s,t) = T_{ladder}(s,t) + \frac{i}{16\pi^2 s} \iint \frac{dt_1\ dt_2}{\sqrt{-\lambda}} \cdot$$

$$\cdot\ T_{ladder}(s,t_1)\ T_{ladder}(s,t_2) + \ldots \qquad (10.24)$$

It is instructive to compare this with eq. (10.1), which yielded the AFS cut. The two expressions for the double ladder exchange are nearly the same, with the sole difference that TT^* in eq. (10.1) is replaced by iTT here. This difference is, however, very important.[16)]

Let us suppose, as before, that T_{ladder} is an even signatured Regge pole:

$$T_{ladder} = \gamma(t)\ s^{\alpha(t)}\ e^{-i\pi\alpha(t)/2} \qquad (10.25)$$

with $\gamma(t)=\gamma_o e^{b_0 t}$ and γ_0 real, and with $\alpha(t)=\alpha_0+\alpha' t$. Then the two-ladder exchange term here gives

$$\frac{\gamma_o{}^2}{32\pi\alpha'}\ e^{b_o t/2}\ \frac{s^{\alpha_c(t)}}{\ell n\ s}\ e^{-\frac{i\pi}{2}\alpha_c(t)}. \qquad (10.26)$$

There is, as in the AFS case, a logarithmic j-plane cut with branch point at $\alpha_c(t)=2\alpha(t/4)-1$. Now, however, the cut occurs in the even signature partial wave amplitude only, and furthermore its discontinuity vanishes at j=0,2,4... It, therefore, introduces no false branch points in t into physical partial wave amplitudes.

Furthermore, if α is near one, so that T_{ladder} is nearly purely imaginary, then the sign of the cut in (10.26) is opposite to that of the AFS cut. For when $T \sim i|T|$, then $iTT^* \sim i|T|^2$, while $iTT \sim -i|T|^2$. Thus the AFS cut interferes constructively with T_{ladder}, while in (10.24) the interference is destructive.[17)]

The first iteration can, as in the AFS case,

be written directly in the j-plane language. We find instead of (10.7) that the double ladder in (10.25) yields

$$\frac{T^{+}(t,j)}{\sin\pi j/2} - i\,\frac{T^{-}(t,j)}{\cos\pi j/2} = \frac{1}{16\pi^{2}}\iint\frac{dt_1\,dt_2}{\sqrt{-\lambda}}\,\cdot$$

$$\cdot\;\frac{\gamma(t_1)\;\gamma(t_2)}{j-\alpha(t_1)-\alpha(t_2)+1}\;e^{\frac{i\pi}{2}(j-\alpha(t_1)-\alpha(t_2)+1)}. \tag{10.27}$$

But now the right-hand side of this equation is real, so that we find no contribution to T^{-}. The cut contributes only in T^{+}; it occurs in the same place as the AFS cut, but it appears only in the even signatured partial wave amplitude.

Next let us continue to positive t by replacing λ by $-\lambda$. We recall that $\gamma(t)=\beta(t)/(\sin\,\pi\alpha(t)/2)$, so we expect $\gamma(t)$ to have poles at positive t when $\alpha(t)=0,2,4\ldots$, namely a right signature integer. Eq. (10.27), therefore, also produces cuts in j with branch points located at $\alpha((\sqrt{t}-m_J)^2)+J-1$, where J is the spin of a particle of mass m_J. These are known as Reggeon-particle cuts,[18] to distinguish them from the cuts with branch points at $2\alpha(t/4)-1$, which are called Reggeon-Reggeon cuts. They are on an unphysical sheet for $t<0$, but one emerges from it to pass through the $m_{J_1}+m_{J_2}$ threshold at J_1+J_2-1 on the physical sheet.

In summary, the asymptotic behavior we find in the ϕ^3 field theory, within the approximations we have made, consists of a Regge pole generated by the ladder graphs corrected by a Regge cut generated by iterated ladders, as in fig. 10.4. If the Regge pole from the ladder is at $\alpha(t)$, then the cut lies at $\alpha_c(t)=2\alpha(t/4)-1$. Thus if we adjust the coupling g so that $\alpha(o)=1$, to produce a constant total cross section, we have $\alpha_c(o)=1$ as well. For $t>0$, the pole lies above the cut; for $t<0$, the cut lies above the pole. At negative t, then, the asymptotic behavior is actually dominated by the cut; exactly at $t=0$, where α and α_c both lie at one, the pole dominates the cut due to the $(\ell n\ s)^{-1}$ factor in the cut.

Should we believe any of these statements are relevant to real hadron physics? There are at least two reasons to be skeptical. First, we have included only a very limited set of graphs. There is every reason to believe that other graphs will modify the results described above, at least in the $t<0$ region. For example, further Regge cuts should be generated by inserting the cut of (10.26) into the eikonal again, and these new cuts should lie above the old ones for $t<0$. There should be a whole hierarchy of cuts, generated by other cuts, and each succeeding cut will lie higher than those which generated it. The ultimate result may be a flat or nearly flat cut for negative t.

A second reason for skepticism is that there

is no reason to believe that ϕ^3 field theory has anything to do with hadrons. As we remarked earlier, the most plausible way for field theory to have relevance to hadron physics is through a quark gluon theory; such a theory is surely not ϕ^3 theory. We should, therefore, proceed, at this point, to discuss other field theories which have more chance of relevance.

11

Massive Quantum Electrodynamics

A field theory of quarks coupled to a massive neutral vector gluon is like quantum electrodynamics with massive photons, and this will be our next subject.

As in the ϕ^3 case, the first step in the analysis of QED consists of summing the most divergent terms as $s \to \infty$ in each order of perturbation theory. As in other types of field theories, in massive electrodynamics this amounts to summing all ladder graphs; however, here it is necessary to include all twisted ladders as well, as indicated in fig. 11.1, in order to preserve gauge invariance. The sum of all ladders plus twisted ladders is called a tower; the amplitude for a tower turns out to be of the form[15)]

$$T_{tower}(s,t) = is \frac{s^a}{(\ln s)^2} F(t). \qquad (11.1)$$

This form is rather similar to what we obtained

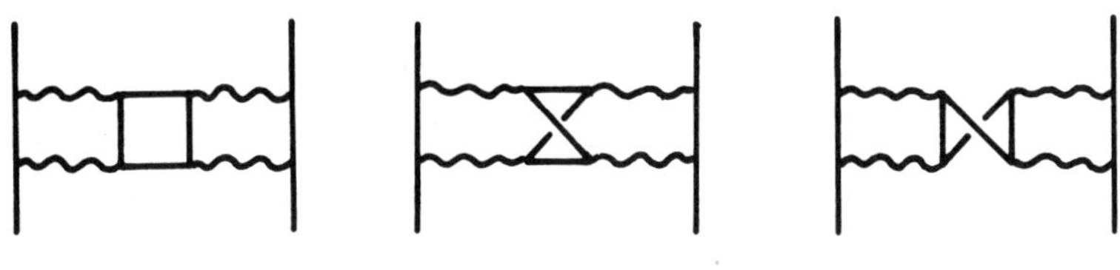

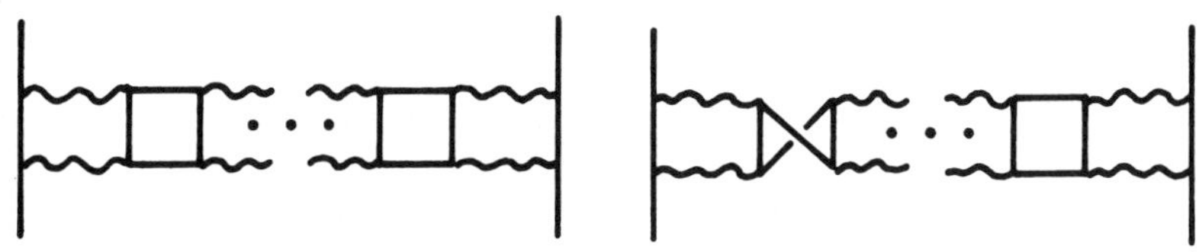

Fig. 11.1. Diagrams included in a "tower" in massive quantum electrodynamics. From ref. 15.

from the same approximation in the ϕ^3 theory. There are, however, some differences which are worth noting.

The first is that in QED, a is a constant, independent of t. In fact, in spinor electrodynamics, $a=(11/32)\pi\alpha^2$ ($\alpha=g^2/4\pi$ where g is the coupling constant) and in scalar electrodynamics, $a=(5/32)\pi\alpha^2$. Thus the value of a is model dependent; but in general it is positive and independent of t.

Second, the tower amplitude is purely imaginary.

Third, the s-dependence is not just a power (which would correspond to a Regge pole) but contains the factor $(1/\ell n\ s)^2$. Thus it represents a fixed soft Regge cut with a $(j-1-a) \ln (j-1-a)$ type of branch point.

Fourth, as in the ϕ^3 theory there is a contribution to the scattering from one particle - in this case vector meson - exchange in the t-channel. This gives rise to a real part of the amplitude proportional to s, which could play a significant role at high energies. It is often argued, however, that such one-particle exchange graphs should not be taken seriously in hadron physics, since they appear in the j-plane as $\delta_{j,1}$ terms and, if the j-plane amplitude is analytic in j, must be absent. To put it another way, they represent the existence of an elementary vector meson, and if such a particle does not exist in nature these terms cannot in fact be present.

The most important feature of the tower is that it violates the Froissart bound. It does this for any value of the coupling constant, since the leading j-plane singularity occurs at $j=1+a$ and a is positive. This is a special case of the remark that any ladder, or tower, in which particles of spin J_1 and J_2 are exchanged has its leading singularity at $j=J_1+J_2-1+O(g^2 \text{ or } g^4)$. For QED the two exchanged particles have spin 1, and $j=1+O(g^4)$. We might recall that in the ϕ^3 theory the leading singularity from a ladder was located at $j=-1+O(g^2)$ (see eq. (9.17)), in accord with the fact that the exchanged particles were spinless. Therefore, in the ϕ^3 case, for small g^2 no violation of the Froissart bound takes place. Only if

g^2 is so large that the term proportional to g^2 in eq. (9.17) is greater than two is the leading singularity to the right of one. Hence in ϕ^3, violation of Froissart by a ladder requires a large coupling constant, so large that one should perhaps be suspicious of the leading ℓn approximation. In QED, in constant, because of the presence of spin one particles, violation takes place for any coupling strength, no matter how weak.

Since the tower violates the Froissart bound, it is impossible for it, by itself, to give the correct asymptotic behavior of the entire amplitude. The tower was what resulted from summing the most divergent graphs in each order; evidently the true asymptotic form must also include graphs which, in a given order, are not the most divergent. The conjecture[15)] is that the additional graphs which are necessary consist of all repeated exchanges of towers, as shown in fig. 11.2. (Note that the end photons of each tower may be mixed up in any manner.) The sum of all these graphs, as we have seen in our eikonal discussion, can be written in the form

$$T(s,t) = 8\pi s \int_0^\infty bdb\ J_o(b\sqrt{-t}) \frac{e^{2i\chi(s,b)} - 1}{2i} \tag{11.2}$$

where

$$T_{tower}(s,t) = 8\pi s \int_0^\infty bdb\ J_o(b\sqrt{-t})\ \chi(s,b). \tag{11.3}$$

Note that T_{tower} is purely imaginary, and therefore

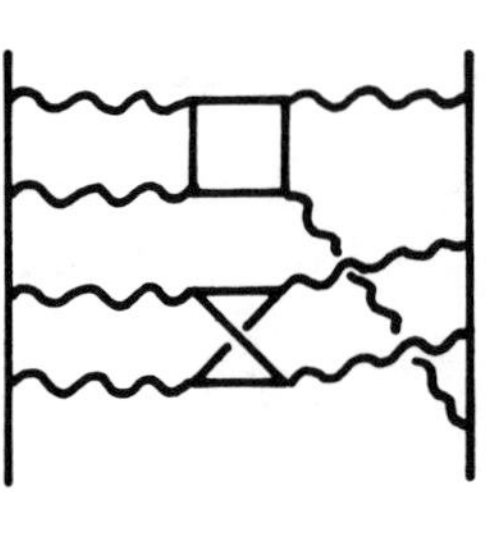

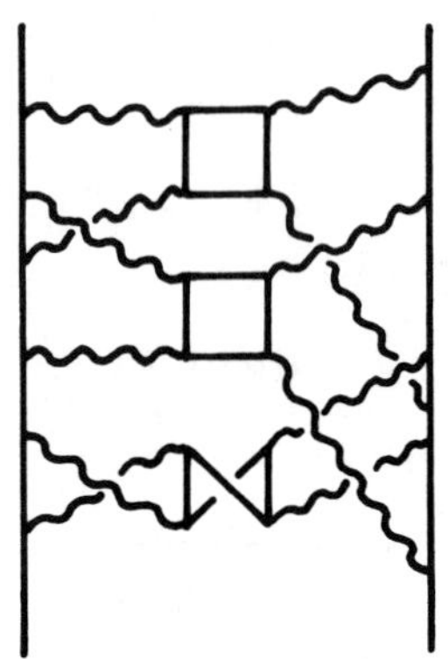

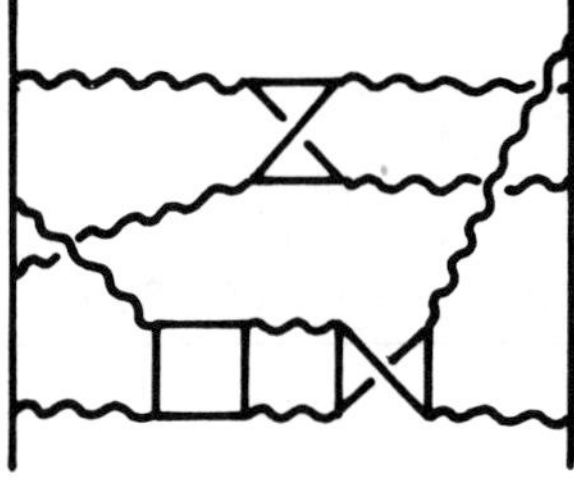

Fig. 11.2. Diagrams showing examples of repeated tower exchange. From ref. 15.

so is χ. We thus have $\delta=0$ and $\chi=-(i/2)\ell n\ \eta$. Hence, the scattering is purely absorptive, and the tower gives the absorption through (11.3). The graph with n exchanged towers is the nth term in the expansion of $e^{2i\chi(s,b)}$ in eq. (11.2).

As $s\to\infty$, $2i\chi\to-\infty$ for a fixed value of b. On the other hand, as $b\to\infty$, $\chi\to0$ for a fixed s. In fact, from eq. (11.3) and the known form for F(t) in eq. (11.1), one finds that $\ell n\ \eta(s,b)$ falls off exponentially in b; we have

$$2i\chi(s,b) = \ell n\ \eta(s,b) \propto -\frac{s^a}{(\ell n\ s)^2}\, e^{-\frac{b}{b_0}}. \quad (11.4)$$

Furthermore, b_0 is a fixed (s independent) number,

since the exponent a in the tower is independent of t. The fact that there is a Yukawa shape, $\exp(-b/b_0)$, in eq. (11.4) is simply a consequence of the existence of a lowest t-channel threshold of fixed mass in the tower. The large b behavior is dominated by the lowest t-channel threshold, and is in general expected to be Yukawa-like.

From eq. (11.2) we find that out to a radius R, $\eta(s,b)-1 \approx -1$ (since $\ell n\ \eta \to \infty$ so $\eta \to 0$) and we have complete absorption. Beyond the radius R, however, $\eta(s,b)-1 \approx 0$, because here $\ell n\,\eta = 0$, or $\eta = 1$. Thus outside of R there is no scattering at all.

The value of R can conveniently be defined as the value of b where $\ell n\ \eta \sim 1$; from (11.4) we see that

$$R(s) \sim b_0\ \ell n \frac{s^a}{(\ell n\ s)^2}\ . \tag{11.5}$$

For sufficiently large s, this becomes more simply $R \sim \ell n\ s$.

The behavior of the cross section is now immediately evident. We have

$$\sigma_T(s) = 4\pi \int_0^R b\,db = 2\pi R^2 \sim \left[\ell n \frac{s^a}{(\ell n\ s)^2}\right]^2 \tag{11.6}$$

so that eventually

$$\sigma_T(s) \sim (\ell n\ s)^2. \tag{11.7}$$

This result reflects the physics: We have a fully absorbing scatterer with a radius R growing logarithmically with the energy.

Away from t=0, the behavior of the amplitude can also be read off from eq. (11.2). We have

$$T(s,t) = 4\pi is\, R(s)\, \frac{J_1(\sqrt{-t}\, R(s))}{\sqrt{-t}} \qquad (11.8)$$

so that

$$T(t,j) = \frac{4\pi R_0^2}{((j-1)^2 - R_0^2 t)^{3/2}} \,. \qquad (11.9)$$

In the j-plane, therefore, the structure is as follows: There are two cuts, with branch points at $j=1\pm i\sqrt{-t}R_0$. For negative t these branch points are complex, at complex conjugate positions, and with constant real part equal to one. For positive t both branch points are real; one is above j=1 at $j=1+R_0\sqrt{t}$ and the other is below at $j=1-R_0\sqrt{t}$. Both cuts, for both signs of t, are hard: They are of the form $(j-\alpha_c)^{-3/2}$.

A few other predictions are,[15] first, that there are zeros in $d\sigma/dt$ at $t_i=-(\beta_i/R)^2$, where β_i are the zeros of J_1. This is obvious from eq. (11.8). Since $R\sim \ln s$, the position of the zeros will move toward t=0 as s increases. There is, therefore, strong shrinkage. Numerically, we have $-t_1\sigma_T(s)=2\pi\beta_1{}^2=35.9$ mb(GeV)^2. For pp scattering, then, where at present energies $\sigma_T\sim 40$ mb, the first zero is near $t \sim -1 (GeV)^2$. It is tempting to identify this with the apparent structure in the present pp data near $t=-1 (GeV)^2$.

The second prediction is also trivial: Since the scattering is from a black disk, we expect $\sigma_{e\ell}/\sigma_T = 1/2$.

Thirdly, we note that the asymptotic amplitude is purely imaginary; thus we predict $ReT/ImT \to 0$ both at $t=0$ and at $t \neq 0$.

Fourth, since the model is based on the dominance of particular classes of graphs, it is explicit enough to make predictions about particle production as well as elastic scattering, and these predictions include a multiplicity of the form $\langle n(s) \rangle \propto s^{a/(1+a)} \ell n s$ and the failure of scaling in inclusive distributions.

These predictions are not all apparent in the existing hadron data, though some look encouraging. Thus the predicted growth of σ_T may be compatible with the pp data from the ISR described in chapter 1, but the prediction that $\sigma_{e\ell}/\sigma_T = 1/2$ is certainly far from confirmation at present.

The discrepancies can perhaps be understood if one believes that, even at the ISR, one is not yet in the asymptotic region. The energy at which the characteristic features of this model sets in depends on the value of a, as is evident from eq. (11.5). As we have seen, a is essentially $(g^2/4\pi)^2$, where g is the neutral vector meson coupling strength. Since the only field theory which could plausibly be considered to have direct relevance to hadron physics is the quark-vector gluon model, g should probably be interpreted as quark-gluon coupling constant. Furthermore, what has actually been calculated is presumably to be identified as the quark-quark scattering cross section.

Detailed fits to recent NAL and ISR pp cross section data have been made using this model,[19] and the assumption that the pp and qq cross sections are essentially the same, at least as far as their s dependence is concerned. The fits require a quite small value of a, and the applicability of the model to hadron physics is, therefore, not free from doubt. Nevertheless, the model and its predictions are extremely interesting, and we must obviously await further data at still higher energies before we can confirm or reject it.

REFERENCES

1. L. Landau, Nucl. Phys. 13,181 (1959).
2. J.C. Polkinghorne, J. Math. Phys. 4, 503 (1963).
   P. Federbush and M. Grisaru, Ann. Phys. 22, 263, 299 (1963).
3. B.W. Lee and R. Sawyer, Phys. Rev. 127, 2266 (1962).
4. Gell-Mann, et al., Phys. Rev. 133, B161 (1964).
5. S. Mandelstam, Phys. Rev. 115, 1741, 1752 (1959).
6. R.J. Eden et al., "The Analytic S-matrix," Cambridge 1966, Section 3.3.
7. H. Bethe and E. Salpeter, Phys. Rev. 84, 1232 (1951).
8. H.W. Wyld, Phys. Rev. D3, 3090 (1971).
9. D. Amati, S. Fubini and A. Stanghellini, Nuovo Cim. 26, 896 (1962).

10. S. Mandelstam, Phys. Rev. 115, 1741 (1959).
11. J.C. Polkinghorne, Phys. Letters 4, 24 (1963).
12. M. Bishari, D. Horn and S. Nussinov. Nucl. Phys. B36, 109 (1972).
13. M. Levy and J. Sucher, Phys. Rev. 186, 1656 (1969).
H.D.I. Abarbanel and C. Itzykson, Phys. Rev. Letters 23, 53 (1969).
S.J. Chang and T.M. Yan, Phys. Rev. Letters 25, 1586 (1970).
G. Tiktopoulos and S. Treiman, Phys. Rev. D3, 1037 (1971).
14. R.J. Glauber, "Lectures in Theoretical Physics," Vol. I, p. 315, Boulder (1958).
15. H. Cheng and T.T. Wu, Phys. Rev. Letters 24, 1456 (1970).
16. J. Finkelstein and M. Jacob, Nuovo Cim. 56A, 681 (1968).
17. The sign of the cut is still in dispute. See, for example,
H.D.I. Abarbanel, Phys. Rev. D6, 2788 (1972).
G.F. Chew, Phys. Rev. D7, 934 (1973).
V. Gribov and A. Migdal, Sov, J. Nucl. Phys. 8, 583, 703 (1969).
I. Muzinich et al., Phys. Rev. D6, 1048 (1972).
18. J. Schwarz, Phys. Rev. 162, 1671 (1967).
19. H. Cheng, J.K. Walker and T.T. Wu, Phys. Letters B44, 97 (1973).

PART IV

MULTIPERIPHERAL MODELS

12

Definition and Properties of Multiperipheral Models

Our examination of field theory suggested the use of the multiperipheral diagram of fig. 12.1 as a zeroth order description of a production amplitude at very high energies.[1] Some of the features of production cross sections associated with this diagram were in qualitative accord with experiment, and some were not. Among the latter was the lack of a sufficiently sharp cutoff in transverse momentum.

In the field theory, for elastic scattering we found that the addition of ladder exchange, and repeated ladder exchange, to the basic one-particle exchange diagram introduced significant new features, such as Regge poles and cuts. It is not unreasonable to expect that

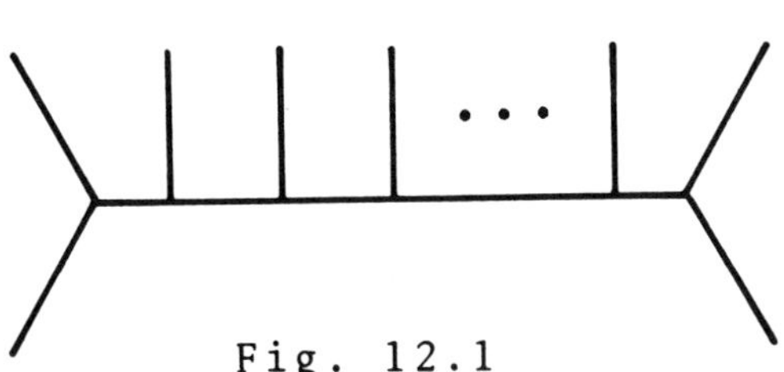

Fig. 12.1

similar modifications to the amplitude of fig. 12.1 will also be important, and may, for example, lead to multi-Regge behavior such as that described in chapter 8. If this is indeed the case, the multiperipheral feature of the amplitude might be preserved, but the exchanged object would not be simply a single particle.[1)]

The idea that production amplitudes can be described by the repeated exchange of some entity or other can also be arrived at in other ways. For example, look at the four-particle production amplitude shown in fig. 12.2a.[*)] When t_2 is near

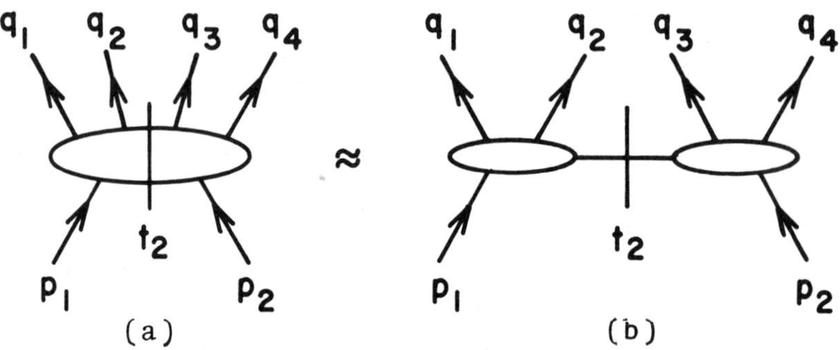

Fig. 12.2

m^2, where m is the mass of a particle permitted to occur in the exchanged channel, the amplitude must factor into two two-particle amplitudes connected by a propagator as shown in fig. 12.2b.

* In the present discussion, where we are not involved with data fitting, we shall generally ignore the distinction between different kinds of particles and shall also suppress spins.

We can write

$$T_{2\to 4}\ (p_1, p_2 \to q_1, q_2, q_3, q_4)$$

$$= T_{2\to 2}(p_1, q_1 + q_2 - p_1 \to q_1, q_2) \cdot$$

$$\cdot \frac{1}{t_2 - m^2}\ T_{2\to 2}(q_3 + q_4 - p_2, p_2 \to q_3, q_4) \quad (12.1)$$

near $t_2 = m^2$, where $t_2 = (p_1 - q_1 - q_2)^2$. It is possible to hope that something like this behavior is valid even if we go to fairly small $t_2 < 0$, though we may have to introduce form factors in t_2, or to include some more complicated off-mass shell behavior in the two-body amplitudes, to describe the variation with t_2.

This argument may be extended to a 2n-particle production amplitude, as shown in fig. 12.3; corresponding to this we would write

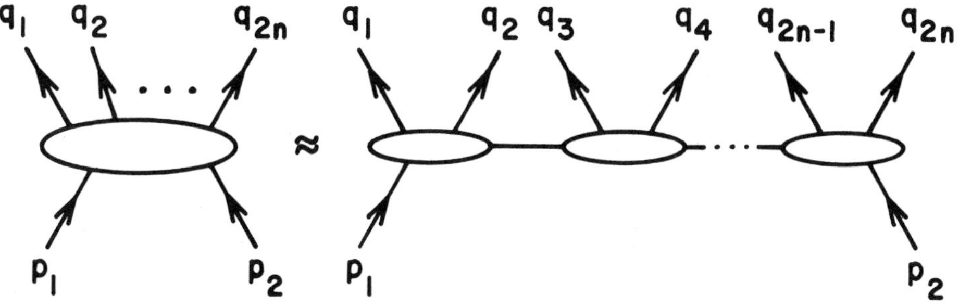

Fig. 12.3

$$T_{2\to 2n}\ (p_1,p_2 \to q_1,\ldots,q_n)$$

$$= T_{2\to 2}(p_1,q_1+q_2-p_1 \to q_1,q_2)\ \frac{1}{t_2-m_2^{\ 2}}\ \cdot$$

$$\cdot T_{2\to 2}(p_1-q_1-q_2,q_3+q_4-p_1+q_1+q_2 \to q_3,q_4)\frac{1}{t_4-m_4^{\ 2}}\cdots$$

$$\cdot\frac{1}{t_{2n-2}-m_{2n-2}^{\ 2}}\ T_{2\to 2}(-p_2+q_{2n}+q_{2n-1},p_2 \to q_{2n-1},q_{2n}) \qquad (12.2)$$

near $t_2=m_2^{\ 2}$, $t_4=m_4^{\ 2}$,... etc. Then we might hope to extend such a form to t_2,t_4...etc. <0 as well, if need be by including off-shell behavior, from factors, and the like.

All of these arguments suggest that we investigate models in which, at small negative t at least, production amplitudes factorize into pieces. There is clearly a large class of such models, differing in what the pieces are supposed to be, but many of the qualitative conclusions to be obtained are common to the entire class. These models are all known as multiperipheral models, and it is these which we shall describe here.

Before proceeding to a detailed discussion of what a multiperipheral model is, and what its predictions are, one additional general comment is appropriate. The crude arguments given above for the validity of a factorized form of production amplitudes are valid only when the various momentum transfers are small. In what follows we shall propose to use these forms over all of phase space, not only in the region where the momentum transfers are small.[2] Evidently, this is possible only if the great majority of production events in fact have small momentum transfers, and experimentally this is true only at high energies. Furthermore, some versions of the multiperipheral model have as the exchanged entity a function whose form can be justified only when the subenergies of produced pairs are large - the MUREX model described in chapter 8 is an example of this. The validity of this class of models then requires either that most events have large subenergies (a circumstance which is experimentally false as evidenced by the fact that, as we saw in chapter 8, MUREX events in the process $\pi^- p \to \pi^+ \pi^- \pi^- p$ at 25 GeV constituted only a small percentage of the total), or that the high subenergy form used is also valid, in some average sense at least, at low subenergies as well. The principle usually invoked to justify this latter situation is duality:[2] For two-body processes the high energy Regge form of the amplitude is also valid at low energies, on the average. It should be kept in mind, however, that the extension of this principle to production amplitudes is by no means obviously true and indeed is quite probably false. For example, duality applies only to the absorptive part of a two-body amplitude, but to justify using multiperipheral models of production in all of phase space it would have to apply to the real part as well.

For these reasons, the arguments "justifying"

some multiperipheral model or other must be viewed with considerable suspicion, and it is by no means clear that they necessarily have any relation to nature at all.

Various kinds of multiperipheral models exist, differing in what and how many objects are repeatedly exchanged. Let us itemize some of them.

The simplest kind of multiperipheral model is one in which only one basic object exists, and in which the production amplitude is simply this object repeated over and over again. Thus

$$\begin{aligned} &T_{2\to n}(p_1,p_2\to q_1,\ldots,q_n) \\ &\quad = \sqrt{\beta_A(t_1)}\; T_o(s_{12},t_1;m_A^2,t_2)\;\cdot \\ &\quad \cdot\; T_o(s_{23},t_2;t_1,t_3,K_1,K_3)\;\cdot \\ &\quad \cdot\; T_o(s_{34},t_3;t_2,t_4,K_2,K_4)\ldots\cdot \\ &\quad \cdot\; T_o(s_{n-1,n},t_{n-1};t_{n-2},m_B^{\,2})\sqrt{\beta_B(t_{n-1})} \end{aligned} \tag{12.3}$$

where the notation has been established previously. The object $T_o(s,t;t_L,t_R,K_L,K_R)$ depends on an energy s and a momentum transfer t, it also depends on the masses t_L and t_R of the left- and right-hand incoming particles, and it, in general, might depend on the variables K_L and K_R defined in the chapter on the multi-Regge theory. $\sqrt{\beta_A}$ and $\sqrt{\beta_B}$

are couplings of the incident particles A and B to the structures T_o at the ends of the multiperipheral chain.

We have already encountered examples of this simplest class of multiperipheral models. The field theory ladder in a ϕ^3 theory, for example corresponds to choosing $T_o = g/(t-\mu^2)$; in this case T_o is independent of s, t_L, t_R and the K's. Another example is the multi-Regge theory. It results from (12.3) with the choice $T_o = \sqrt{V(t_L,t,K_L)} \cdot s^{\alpha(t)} \xi(t) \sqrt{V(t,t_R,K_R)}$, as in the natural generalization of eq. (8.4).

A modification of (12.3), which is indistinguishable from it when all subenergies are large, and which is sometimes computationally more convenient, is to replace the variable $s_{i,i+1}$ in each T_o by $\frac{s}{s_i \bar{s}_i} m_T^4$, where $s_i = (q_1 + .. + q_i)^2$ and $\bar{s}_i = (q_{i+1} + ... + q_n)^2$, and where m_T is an "average" transverse mass.

Another modification of eq. (12.3) which is frequently discussed is to say there are a number of basic objects T_i, rather than just one, and that any of these can be exchanged in a given multiperipheral leg. Thus (12.3) would be replaced by

$$T_{2\to n} = \sum_{i_1 .. i_{n-1}} \sqrt{\beta_A^{i_1}}\, T_{i_1} T_{i_2} \cdots T_{i_{n-1}} \sqrt{\beta_B^{i_{n-1}}}. \tag{12.4}$$

As obvious example of this situation is a field

theory with a number of different scalar particles present; this would lead to (12.4) with $T_i = g_i/(t-\mu_i^2)$ in the ladder approximation, provided that the coupling constant factored: that is, provided that $g_{ij} = \sqrt{g_i}\sqrt{g_j}$ is the coupling of the produced particle and the ith and j^{th} exchanged particles. Another example is the multi-Regge model with several trajectories, and with factorized couplings.

In some situations models in which the production amplitude has an alternating structure, as in fig. 12.3, are attractive. We could write

$$
\begin{aligned}
T_{2\to n}(p_1, p_2 \to q_1, \ldots, q_n) = {} & T_o(s_{12}, t_1; m_A^2, t_2, K_2) \cdot \\
& \cdot X(s_{23}, t_2; t_1, t_3, K_1, K_3) \cdot \\
& \cdot T_o(s_{34}, t_3; t_2, t_4, K_2, K_4) \ldots \cdot \\
& \cdot X(s_{n-2,n-1}, t_{n-2}; t_{n-3}, t_{n-1}, K_{n-3}, K_{n-1}) \cdot \\
& \cdot T_o(s_{n-1,n}, t_{n-1}; t_{n-2}, m_B^2, K_{n-2})
\end{aligned}
\tag{12.5}
$$

in terms of two basic entities T_o and X. An illustration of this would be the production amplitude near $t_2 = m^2$, $t_4 = m^2$....etc. where we would have T_o the two-body amplitude itself, and X would be a propagator, as in eq. (12.2):

$$T_o(s,t;t_L,t_R,K_L,K_R) = T(s,t),$$

$$X(s,t;t_L,t_R,K_L,K_R) = \frac{1}{t-m^2}\ . \qquad (12.6)$$

With this interpretation, we note, n must be even.

Modifications of such models are possible, too. For example, one could again say that the function X depends not on a subenergy $s_{i,i+1}$ but rather on the combination $\frac{s}{s_i \bar{s}_i} \cdot m_T^4$. Other special cases which may be of interest are those in which X, say, depends only on momentum transfer, or is independent of t_L, t_R, K_L and K_R, and so forth.

The crucial common ingredient in all of these versions of the multiperipheral model is the repetitive nature of the production amplitude. The fact that as n increases, no new structures appear, and that correlations among the produced particles extend only a finite distance down the multiperipheral chain, leads to several striking qualitative conclusions. These conclusions are evidently common to all versions of the multiperipheral model, and it is appropriate to describe some of them here.

First, the functions T_o, or T_o and X, can evidently be chosen to have a rapid falloff in momentum transfer. This will then automatically provide a sharp cutoff in q_T in the production amplitude. (Note, though, that this is an input, not a consequence, of the multiperipheral model.)

Second, since a given q_i is related only to q_{i-1} or q_{i+1} through T_o and/or X, short-range correlations are implied. The multiplicity, also, is logarithmically dependent on the energy.

Third, all the Mueller analysis, including the results of limiting fragmentation and scaling, can be made to follow from the multiperipheral model.

Fourth, if T_o is chosen to have Regge asymptotic behavior, then MUREX follows.

The repetitive nature of the production amplitude also permits us, through the intervention of the s-channel unitarity relation, to construct an integral equation for the elastic scattering amplitude - or, rather, for its s-channel absorptive part.[3] Let us illustrate how this comes about in detail with a particular version of the multiperipheral model, namely that in which we take

$$\begin{aligned} & T_{2\to n}(p_1, p_2 \to q_1, \ldots, q_n) \\ & \quad = T_o(s_{12}, t_1; m_A{}^2, t_2)\, X(t_2) \cdot \\ & \quad \cdot\, T_o(s_{34}, t_3; t_2, t_4)\, X(t_4) \ldots X(t_{n-2}) \cdot \\ & \quad \cdot\, T_o(s_{n-1,n}, t_{n-1}; t_{n-2}, m_B{}^2). \end{aligned} \tag{12.7}$$

We must at once make the assumption that when two such production amplitudes, one for $p_1+p_2 \to q_1+\ldots+q_n$, the other for $p_1'+p_2' \to q_1 \ldots +q_n$, are inserted into the s-channel unitarity relation, the ordering of the particles $q_1 \ldots q_n$ is the same in both amplitudes. That is, we must discard the possibility that lines cross. The argument usually given to justify this is that if T_o and X cut off sharply in t, and since graphs with crossed lines involve larger t values than graphs without crossed lines, the former are damped. But this is really true only if the subenergies are large, and in using the s-channel unitarity relation we must integrate over all of phase space, including the region where the subenergies are small. Furthermore, even if a single crossed line graph is small, there are many of them and it is by no means clear that the sum of all of them is small.[4)] In summary, the neglect of crossed lines may well not be justified, and this should be kept strongly in mind in assessing the relevance of all predictions of the multiperipheral model.

With this proviso, let us insert eq. (12.7) into the s-channel unitarity relation, and separate out the n=2 term from the sum over intermediate states. The n=2 term we shall call $A_2(s,t)$; it is given by

$$A_2(s,t) = \frac{1}{16\pi^2 s}\iint \frac{dt_1\ dt_1'}{\sqrt{-\lambda(t,t_1,t_1')}} \cdot$$

$$\cdot T_o(s,t_1,m_A^2,m_B^2)T_o^*(s,t_1',m_A^2,m_B^2). \tag{12.8}$$

The entire unitarity relation then states that

$$A(s,t) = A_2(s,t)+\sum_{n=4}^{\infty} \frac{1}{2} \prod_{i=1}^{n}\int\frac{d^3q_i}{(2\pi)^3} \frac{1}{2\omega_i}(2\pi)^4 \cdot$$

$$\cdot\ \delta^{(4)}(P-\sum_{j=1}^{n} q_j)\ T_o(s_{12},t_1;m_A^2,t_2) \cdot$$

$$\cdot\ X(t_2)\ldots X(t_{n-1})T_o(s_{n-1,n},t_{n-1};t_{n-2},m_B^2)\cdot$$

$$\cdot\ T_o^*(s_{12},t_1';m_A^2,t_2')X^*(t_2')\ldots X^*(t_{n-1}')\cdot$$

$$\cdot\ T_o^*(s_{n-1,n},t_{n-1}';t_{n-2}',m_B^2) \tag{12.9}$$

where, as before, A(s,t) is the absorbtive part of the elastic amplitude, and where $t_i=(p_1-q_1\ldots-q_i)^2$ and $t_i'=(p_1'-q_1-\ldots-q_i)^2$. We recall that in eq. (12.7), n is even for the specific model under discussion. Hence the sum in (12.9) begins at n=4, not n=3.

Now let us insert into the second term on

the right-hand side of eq. (12.9) the expression

$$1 = \int \frac{d^4Q}{(2\pi)^4} (2\pi)^4 \delta^{(4)}(Q-q_1-q_2)$$

and let us also extend the definition of A(s,t) to include off-shell values of the external masses. We may then rewrite (12.9) as

$$A(s,t) = A_2(s,t) + \int \frac{d^4Q}{(2\pi)^4} \int \frac{d^3q_1}{(2\pi)^3} \int \frac{d^3q_2}{(2\pi)^3} \frac{(2\pi)^4}{4\omega_1\omega_2} \cdot$$

$$\cdot \delta^{(4)}(Q-q_1-q_2) \; T_o(s_{12},t_1,m_A{}^2,t_2)X(t_2) \cdot$$

$$\cdot A((P-Q)^2,t;t_2,t_2{}') \; T_o{}^*(s_{12},t_1{}';m_A{}^2,t_2{}') \cdot$$

$$\cdot X^*(t_2{}'). \qquad (12.10)$$

The evaluation of the integrals on d^3q_1 and d^3q_2 is easy, and, with an extension of the definition (12.8) to off shell external masses, yields

$$A(s,t) = A_2(s,t) + 2 \int \frac{d^4Q}{(2\pi)^4} A_2(Q^2,t;t_2,t_2{}') \cdot$$

$$\cdot X(t_2)X^*(t_2{}')A((P-Q)^2,t;t_2,t_2{}') \qquad (12.11)$$

where $t_2=(p_1-Q)^2$ and $t_2'=(p_1'-Q)^2$.

Eq. (12.11) is not an integral equation because of the off-mass shell dependence in A and A_2 inside the integral, but it can easily be made to be one by letting all legs on all A's and A_2's be off-shell. That obviously yields

$$A(s,t;t_L,t_L',t_R,t_R') = A_2(s,t;t_L,t_L',t_R,t_R')$$

$$+ 2\int\frac{d^4Q}{(2\pi)^4}\, A_2(Q^2,t;t_L,t_L',t_M,t_M')\cdot$$

$$\cdot X(t_M)X^*(t_M')A((P-Q)^2,t;t_M,t_M',t_R,t_R') \tag{12.12}$$

where $t_M=(p_1-Q)^2$, $t_M'=(p_1'-Q)^2$. This equation is to be coupled with the extension of eq. (12.8) to off-shell values, namely

$$A_2(s,t;t_L,t_L',t_R,t_R') =$$

$$= \frac{1}{16\pi^2 s}\iint\frac{dt_M\, dt_M'}{\sqrt{-\lambda(t,t_M,t_M')}}\cdot$$

$$\cdot T_o(s,t_M;t_L,t_R)T_o^*(s,t_M';t_L',t_R'). \tag{12.13}$$

Eq. (12.12) is, of course, simply a Bethe-Salpeter equation with inhomogeneous term A_2 and Kernel A_2XX^*; it and eq. (12.13) are represented by the diagrammatic equation shown in fig. 12.4.

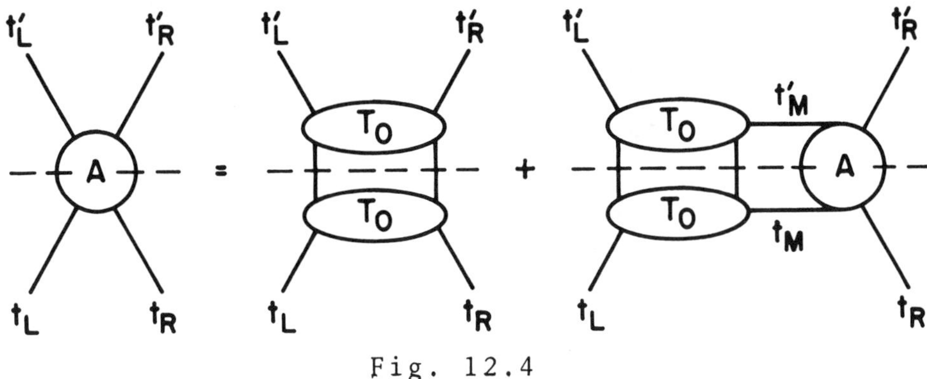

Fig. 12.4

It is a generalization of the Bethe-Salpeter equation given in our discussion of the field theoretic ladder of the ϕ^3 theory to a case in which the propagators of the ϕ^3 theory are replaced alternately by T_o and X.

It will be convenient to change the variables of integration in eq. (12.12) from d^4Q to those on which A and A_2 directly depend, namely $s_1=Q^2$, $s_2=(P-Q)^2$, t_M and t_M'. We find by a straightforward calculation that (12.12) can be replaced by

$$A(s,t;t_L,t_L',t_R,t_R')=A_2(s,t;t_L,t_L',t_R,t_R')$$

$$+\frac{1}{(2\pi)^4 s}\iint\frac{ds_1 ds_2 dt_M dt_M'}{\sqrt{-\lambda(t,t_M,t_M',s,s_1,s_2)}}\cdot$$

$$\cdot A_2(s_1,t;t_L,t_L',t_M,t_M')\;X(t_M)\;X^*(t_M')\cdot$$

$$\cdot A(s_2,t;t_M,t_M',t_R,t_R') \qquad (12.14)$$

where, for $s>>s_1,s_2$, $\lambda\to\lambda(t,t_M,t_M')$ (see appendix E). For a strongly damped $X(t_M)$ the only important region of integration is $s>s_1s_2$. We shall henceforth use this restriction.

Eq. (12.14) is a very basic one, and it is perhaps worthwhile to list its equivalents in some of the other versions of the multiperipheral model which we have mentioned.

(i) In the event that T_o is independent of off-shell masses; that is, if $T_o(s,t;t_L,t_R)=T_o(s,t)$ only, then the off-shell dependence in (12.14) can be dropped and we have

$$A(s,t) = A_2(s,t) + \frac{K(t)}{\pi^2 s}\iint ds_1 ds_2\;\cdot$$

$$\cdot A_2(s_1,t)\;A(s_2,t) \qquad (12.15)$$

where

$$K(t) = \frac{1}{16\pi^2}\iint \frac{dt_M\, dt_M'}{\sqrt{-\lambda(t,t_M,t_M')}}\, X(t_M)\, X^*(t_M'). \tag{12.16}$$

This is, obviously, a far simpler equation than (12.14) and, as we shall see shortly, it can be solved exactly analytically.

(ii) A more complicated example of (12.14) results if X is permitted to have some s dependence. For example, we could replace X(t) in (12.14) by $X(s/s_1 s_2, t)$; this corresponds to a production amplitude in which the $X(t_i)$ in eq. (12.7) is replaced by $X(s/s_i\bar{s}_i, t_i)$.

(iii) Another case of interest is that in which the production amplitude involves only a single repeated entity T_o, rather than two, T_o and X. For this case a repetition of the arithmetic which led to (12.14) yields in the strong ordering approximation

$$\begin{aligned} &A(s,t;t_L,t_L',t_R,t_R') = \\ &\quad = \beta_A(t_L,t_L')\, A_2(s,t;t_L,t_L',t_R,t_R') \cdot \\ &\quad \cdot \beta_B(t_R,t_R') + \frac{1}{16\pi^3 s}\iiint \frac{ds_1\, dt_M dt_M'}{\sqrt{-\lambda}} \cdot \\ &\quad \cdot T_o\left(\frac{s}{s_1}m_T^2, t_M; t_L, -m_T^2\right) T_o^*\left(\frac{s}{s_1}m_T^2, t_M'; t_L', -m_T^2\right) \\ &\quad \cdot A(s_1,t;t_M,t_M',t_R,t_R'). \end{aligned} \tag{12.17}$$

(iv) In the special case that the single object T_o is independent of off-shell masses, (12.17) reduces to the convenient equation

$$A(s,t) = \beta_A\, A_2(s,t)\, \beta_B + \frac{1}{\pi}\int^{s} \frac{ds_1}{s_1}\,\cdot$$

$$\cdot A_2\left(\frac{s}{s_1}\, m_T^{\,2}, t\right) A(s_1,t). \qquad (12.18)$$

Having commented on other versions, let us now return to eq. (12.14) and manipulate it further. A useful transformation of (12.14) consists of projecting it into t-channel partial wave amplitudes. We may use the standard definition given in appendix D, namely that at large s

$$A(s,t) = \frac{1}{2\pi i}\int_{c-i\infty}^{c+i\infty} dj\; s^{j}\; T(t,j) \qquad (12.19)$$

where $T(t,j)=T^{(+)}(t,j)+T^{(-)}(t,j)$ is the t-channel partial wave amplitude, to integrate (12.14) over the region $s>s_1s_2$ and obtain the result

$$T(t,j;t_L,t_L',t_R,t_R') = T_2(t,j;t_L,t_L',t_R,t_R')$$

$$+\frac{1}{j+1}\,\frac{1}{(2\pi)^4}\iint \frac{dt_M\, dt_M'}{\sqrt{-\lambda(t,t_M,t_M')}}\, T_2(t,j;t_L,T_L',t_M,t_M')\cdot$$

$$\cdot X(t_M)X^{*}(t_M')T(t,j;t_M,t_M',t_R,t_R'). \qquad (12.20)$$

In writing (12.20), we have used expressions analogous to (12.19) for the off-shell amplitude and for A_2 as well as for A. The special case,(12.15), in which the off-shell dependence was suppressed, replaces (12.20) by the algebraic equation

$$T(t,j) = \beta_A\, T_2(t,j)\beta_B + \frac{1}{\pi^2}\, T_2(t,j)\cdot$$

$$\cdot\frac{K(t)}{j+1}\, T(t,j) \tag{12.21}$$

with the solution

$$T(t,j) = \frac{\beta_A\, T_2(t,j)\ \beta_B}{1 - \frac{1}{\pi^2}\frac{K(t)}{j+1}\, T_2(t,j)}\,. \tag{12.22}$$

The more general case in which X has an s dependence, but still without off-shell dependence in T_o, replaces $\frac{K(t)}{j+1}$ in (12.22) by $K(t,j)$, which is the j-plane projection of

$$\frac{1}{s}K(s,t) = \frac{1}{16\pi^2 s}\iint\frac{dt_M dt_M'}{\sqrt{-\lambda}}X(s,t_M)X^*(s,t_M')\,. \tag{12.23}$$

Finally, the multiperipheral model containing only a single structure, leading to (12.18), gives

$$T(t,j) = \beta_A\beta_B\ T_2(t,j)/\left[1-\frac{1}{\pi}\ T_2(t,j)\right]. \quad (12.24)$$

We shall find these simple versions of the multiperipheral equation very helpful in understanding what the physical content of multiperipheral models really is. The simple cases have all the general features of the more complicated versions, and they are far more transparent; the additional information obtained by attempting to actually solve integral equations like (12.20) is usually not worth the trouble.

For most reasonable choices of T_o and X, eq. (12.20) - and indeed other versions of the multiperipheral equation as well - is a Fredholm equation. This is true, we should emphasize, only if T_o and X are given functions. If T_o, for example, is itself an unknown expressed in some way or other in terms of T, as we will see is the case in bootstrapped versions of the multiperipheral model, then eq. (12.20) is no longer linear and is, therefore, definitely not a Fredholm equation.

When the equation is a Fredholm equation, however, some general conclusions about the solution can be drawn without solving the equation in detail.[5)] In particular, the leading j-plane singularity must be a moving Regge pole, and furthermore the pole residue must be factorizable. The position of the pole and its residue are real below t-channel threshold, and above values of t at which it collides with a branch point or other pole. These conclusions follow only from assuming that the input T_o has Regge asymptotic behavior.

13

Solutions of the Multiperipheral Equations

We have now carried the arithmetic of the multiperipheral model as far as we can without becoming specific about what the basic structures T_o and X are. So let us at this point select the simplest formula we have, and see what it gives us for various choices of T_o. We shall suppose that the inelastic production amplitude has the ultrasimple form

$$T_{2\to n}(p_1, p_2 \to q_1, \ldots, q_n)$$

$$= \sqrt{\beta_A}\; T_o(s_{12}, t_1)\; T_o(s_{23}, t_2) \ldots\ldots$$

$$\cdot\; T_o(s_{n-1,n}, t_{n-1})\; \sqrt{\beta_B} \qquad (13.1)$$

which, as we have seen, results approximately in

$$T(t,j) = \frac{\beta_A \beta_B \; T_2(t,j)}{1-\frac{1}{\pi}\, T_2(t,j)} \tag{13.2}$$

where $T_2(t,j)$ is the Mellin transform of $A_2(s,t)$, and

$$A_2(s,t) = \frac{1}{16\pi^2 s}\iint \frac{dt_1 dt_1'}{\sqrt{-\lambda}}\; T_o(s,t_1) T_o^*(s,t_1'). \tag{13.3}$$

The simplest thing we can choose for $T_o(s,t)$ is a fixed Regge pole:*)

$$T_o(s,t) = \beta_o(t)\; s^{\alpha_o}. \tag{13.4}$$

With this choice, we find[2)]

$$A_2(s,t) = \beta(t) s^{2\alpha_o - 1} \tag{13.5}$$

where

$$\beta(t) = \frac{1}{16\pi^2}\iint \frac{dt_1\; dt_1'}{\sqrt{-\lambda}}\; \beta_o(t_1)\beta_o(t_1'). \tag{13.6}$$

---

* Note that in this case we do not mean (13.1) to apply to the elastic amplitude $T_{2\to 2}$; we shall discuss later (in chapter 14) versions of the multiperipheral model in which $T_{2\to 2}$ is included in the basic form (13.1).

The j-plane projection corresponding to this A_2 is simply

$$T_2(t,j) = \frac{\beta(t)}{j-(2\alpha_o-1)} \tag{13.7}$$

thus we obtain, from (13.2),

$$T(t,j) = \frac{\beta(t)\beta_A\beta_B}{j-\alpha(t)} \tag{13.8}$$

where

$$\alpha(t) = 2\alpha_o - 1 + \frac{1}{\pi}\beta(t). \tag{13.9}$$

We note that this $\alpha(t)$ can be linear in t only over a limited range of t, for any reasonable input $\beta_o(t)$. Going back to the s,t language then produces our answer: with this input, the multiperipheral model gives a moving Regge pole:

$$A(s,t) = \beta_A\beta_B\beta(t)\ s^{\alpha(t)} \tag{13.10}$$

Several comments are in order.

(i) In the special case with $\alpha_o=0$, the output Regge pole is $\alpha(t)=-1+\frac{1}{\pi}\beta(t)$, a result reminiscent of what we found in field theory from a ladder. We note that $\alpha_o=0$ corresponds to $T_o(s,t)=\beta(t)$, of which a propagator $g/(t-\mu^2)$ is a special case. The fact that the resulting $\alpha(t)$ is the same is, therefore, not surprising; we should also remark that $\frac{1}{\pi}\beta(t)$ here coincides with the function $g^4 f(t)$ we found in the field theory case with the propagator as input.

(ii) The inelastic cross section associated with (13.10) is

$$\sigma_I(s) = \beta_A \beta_B \beta(o)\, s^{2\alpha_o - 2 + \frac{1}{\pi}\beta(o)}. \tag{13.11}$$

It is, therefore, a constant provided that

$$\alpha_o = 1 - \frac{1}{2\pi}\, \beta(o). \tag{13.12}$$

We note from (13.6) that

$$\beta(o) = \frac{1}{16\pi} \int_{-\infty}^{0} dt\, (\beta_o(t))^2 \tag{13.13}$$

is positive definite; hence to produce a constant inelastic cross section we must have $\alpha_o < 1$. For a given $\alpha_o, \beta(o)$ is determined by (13.12). The value of the cross section is then also determined by (13.11). For example, if $\alpha_o = 1/2$ (a popular value) we have $\beta(o) = \pi$ whereupon $\sigma_I = \beta_A \beta_B \pi (\mathrm{GeV})^{-2}$. In the ϕ^3 field theory, for example, where $\beta_A = \beta_B = g$ and where $\beta_o(t) = g/t - \mu^2$, we can calculate g from (13.13) and the requirement that $\beta(o) = \pi$. We find $g^2/16\pi\mu^2 = \pi$, and this in turn yields

$\sigma_I = 16\pi^3/\mu^2$. For any reasonable value of μ this is an immensely large cross section. Thus we conclude that, in this model, the coupling strength required to yield a constant cross section is much stronger than the couplings which seem to occur in nature.

(iii) The contribution to $T(t,j)$ from the n-particle intermediate state is

$$T_n(t,j) = \beta_A \beta_B \; T_2(t,j) \; (\tfrac{1}{\pi} \; T_2(t,j))^{n-2} \tag{13.14}$$

as can be seen directly from (13.2). Hence with our fixed Regge pole input

$$T_n(t,j) = \beta_A \beta_B \beta(t) \; \frac{(\frac{1}{\pi} \; \beta(t))^{n-2}}{(j-(2\alpha_o-1))^{n-1}} \tag{13.15}$$

and therefore

$$A_n(s,t) = \beta_A \beta_B \beta(t) \; s^{2\alpha_o-1} \; \frac{(\frac{1}{\pi} \; \beta(t) \ell n \; s)^{n-2}}{(n-2)!}. \tag{13.16}$$

Consequently the partial cross sections are

$$\sigma_n(s) = \beta_A \beta_B \beta(o) \; s^{2\alpha_o-2} \; \frac{(\frac{1}{\pi} \; \beta(o) \ell n \; s)^{n-2}}{(n-2)!}. \tag{13.17}$$

They are Poisson distributed, and vanish with a power of s (since $\alpha_o < 1$); yet they add together to give a constant inelastic cross section if (13.12) is satisfied. Furthermore, it is evident from (13.17) that

$$<n(s)> = \frac{1}{\pi} \beta(o) \ln s + 2 \tag{13.18}$$

(iv) We can also make a connection with the generating function formalism developed in chapter 3. We note that

$$Q(s,z) = \frac{1}{\sigma} \sum_n z^n \sigma_n = \frac{\sum_n z^n A_n(s,o)}{s\sigma}$$

$$= \frac{1}{s\sigma} A(s,o) \Big|_{\beta_o \to \sqrt{z}\beta_o} \tag{13.19}$$

where, in the last step, we have multiplied each coupling constant (including β_A and β_B) which is implicit in A by a factor of $\sqrt{z}$. This corresponds to changing $\beta(o) \to z\beta(o)$ and, therefore, eq. (13.10) leads to

$$Q(s,z) = z^2 \, s^{\frac{1}{\pi}\beta(o)(z-1)} . \tag{13.20}$$

From here it is straightforward to conclude that

$$\langle n\rangle = \frac{\partial}{\partial z} \ln Q(z)\Big|_{z=1} = 2 + \frac{1}{\pi}\beta(o) \ln s$$

which coincides, of course, with (13.18). Furthermore,

$$f_k = \frac{\partial^k}{\partial z^k} \ln Q(z)\Big|_{z=1} = (-)^{k-1}\, 2(k-1)! \qquad k \geq 2 \tag{13.21}$$

which is a statement of short-range correlations. As a matter of fact, if we were to define $Q(z)=\frac{1}{\sigma}\Sigma z^{n-2}\sigma_n$ we would find that all correlation parameters are zero, since (13.17) is a Poisson distribution in n-2.

(v) We note that the production amplitude $T_{2\to n}$ is proportional to $s_{12}{}^{\alpha_o} s_{23}{}^{\alpha_o} s_{34}{}^{\alpha_o} \ldots s_{n-1,n}{}^{\alpha_o}$, and, therefore, peaks when the subenergies $s_{i,i+1}$ are all equal and equal to $(s)^{1/n-1}$. We recall from our discussion of the kinematics of production in chapter 8 that large subenergies mean strong ordering: ω_i/ω_{i+1} large, or large rapidity separation: $y_i - y_{i+1}$ large. Thus the preferred configuration for production is when all produced particles are evenly spaced on a rapidity plot. All this can be made explicit by calculating the distribution in rapidity of the ith produced particle in an n-particle production process.[6] The cross section is

$$\frac{d\sigma_{n,i}}{dy} \propto s^{2\alpha-2} \frac{y^{i-1}}{(i-1)!} \frac{(Y-y)^{n-i-2}}{(n-i-2)!}. \tag{13.22}$$

This distribution is shown in fig. 13.1.

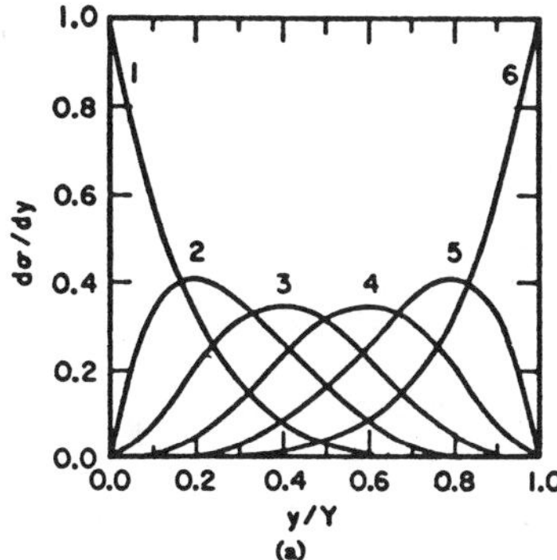

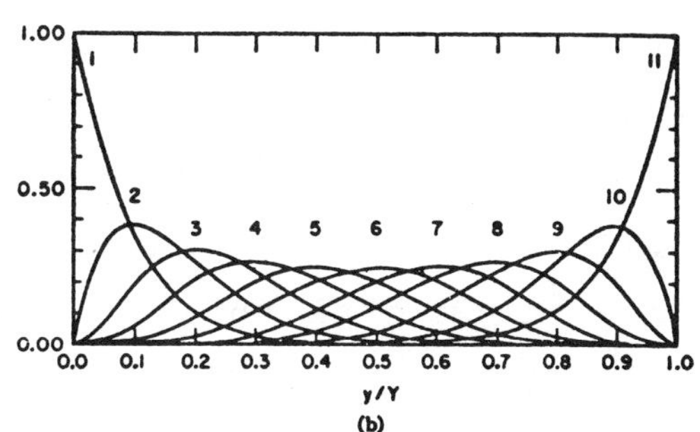

Fig. 13.1. Distribution in longitudinal momentum of the ith produced particle in a simplified Chew-Pignotti model; $d\sigma_i/dy$ for the particle is given in arbitrary units. Distributions are shown for (a) six and (b) eleven produced particles. Taken from ref. 6.

(vi) It is evident that this tendency to space the produced particles evenly in a rapidity plot will result in a flat rapidity distribution for the one-particle inclusive cross section.[6] As can be seen from fig. 13.1, the distributions for each i, when summed over, lead to flat plateau. We have

$$\frac{d\sigma}{dy} = \sum_{n,i} \frac{d\sigma_{n,i}}{dy} \tag{13.23}$$

and this yields a rectangle in shape. The sharp corners on the distributions, producing a very odd shape in the fragmentation regions, are obviously the result of the extreme simplicity of the model and are not to be taken seriously; nevertheless, scaling and limiting fragmentation (albeit with a funny shape) are consequences of this and indeed of any multiperipheral model.[7)]

(vii) There is automatically a cutoff in the q_T of the produced particles if we assume the input $\beta_0(t)$ to cut off in momentum transfer t.[8)] The production amplitude $T_{2\to n}$ maximizes when all $q_T^{\,2}$ are zero.

To summarize, at this most elementary level, the multiperipheral model with a fixed Regge pole input gives a moving Regge pole output. The output pole can be adjusted to pass through j=1 at t=0, although numerically the values of α_0 and $\beta_0(t)$ necessary to arrange this do not provide a very realistic value of the cross section. In this picture, then, the Pomeron is to be viewed as generated by lower-lying trajectories (since $\alpha_0<1$); if these lower-lying trajectories are fixed, the Pomeron is a simple moving pole.

An obvious difficulty with this version of the world is this: If the input in the multiperipheral model consists of Regge poles, and if these generate an output Regge pole, why is the output pole not itself included in the input?

We could try to make the output pole and

the input pole coincide, but since the output pole moves while the input pole does not, this is evidently not possible in general. At best we could make them coincide at t=0; if we do this, then we must require

$$\alpha_o = \alpha(o) = 2\alpha_o - 1 + \frac{1}{\pi}\beta(o) \qquad (13.24)$$

so that

$$\alpha_o = 1 - \frac{1}{\pi}\beta(o) < 0.$$

But then $\alpha(o)<1$, so that $\sigma_T(s)$ vanishes like a power of s: The Pomeron is a pole lying below one at t=0. We shall return to this feature in much more detail later.

Let us next make the input slightly more complicated, and see what changes this produces.[9] Suppose we take for T_o a moving Regge pole: $T_o(s,t)=\beta_o(t)s^{\alpha_o(t)}$.

As we have seen before, in the discussion of field theory, with a moving Regge pole input $A_2(s,t)$ will contain the AFS cut. This is a logarithmic j-plane singularity, located at $\alpha_c=2\alpha(t/4)-1$. We will, therefore, have

$$T_2(t,j) = \gamma(t)\,\ell n(j-\alpha_c(t)) \qquad (13.25)$$

where $\gamma(t)$ is quadratically related to $\beta_o(t)$ and may roughly speaking be taken as a constant for small t. Consequently, from (13.2) we find

$$T(t,j) = \frac{\beta_A \beta_B \gamma(t) \ \ell n(j-\alpha_c(t))}{1-\frac{1}{\pi} \gamma(t) \ \ell n(j-\alpha_c(t))} . \tag{13.26}$$

Thus T also has the AFS branch point, but T is no longer infinite at $j=\alpha_c(t)$; it has become finite there. Thus the character of the branch point has changed.

It is also of interest to discuss situations where the input $T_o(s,t)$ is the sum of several different parts. Suppose, for example, that T_o contains two Regge poles:

$$T_o(s,t) = \beta_1(t) \ s^{\alpha_1(t)} + \beta_2(t) \ s^{\alpha_2(t)} .$$

Then $T_2(t,j)$ contains three branch points, at $\alpha_{11}(t)=2\alpha_1(t/4)-1$, $\alpha_{12}(t)=\alpha_1(t/4)+\alpha_2(t/4)-1$, and at $\alpha_{22}(t)=2\alpha_2(t/4)-1$, all of which are logarithmic. Thus we find, in analogy to (13.26), that

$$T_2(t,j)T^{-1}(t,j) \propto 1-\frac{1}{\pi} \gamma_{11} \ \ell n(j-\alpha_{11})$$

$$-\frac{1}{\pi} \gamma_{12} \ \ell n(j-\alpha_{12})-\frac{1}{\pi} \gamma_{22} \ \ell n(j-\alpha_{22}) . \tag{13.27}$$

A special case of this is the situation when one input pole, say α_1, is flat and the other moves: Then T_2 has a pole at $j=2\alpha_1-1$, a flat cut at $\alpha_{12}=\alpha_1+\alpha_2(o)-1$, as well as the usual AFS cut at $j=\alpha_{22}(t)$. Now we find

$$T_2(t,j)T^{-1}(t,j) \propto j-(2\alpha_1-1)-\frac{1}{\pi}\,\beta_{11}(t)$$

$$-\frac{1}{\pi}\,\gamma_{12}\,\ell n(j-\alpha_{12})-\frac{1}{\pi}\,\gamma_{22}\,\ell n(j-\alpha_{22}) \quad (13.28)$$

and in this expression γ_{12} and γ_{22} are proportional to $j-(2\alpha_1-1)$.

We can proceed further, to more and more complicated inputs. For each type of input so far studied, the output turns out to be different, and, generally, much more complicated. The following table summarizes the situation.

Input	Output
Flat Pole	Moving Pole
Moving Pole	Moving Cut
Several Poles	Moving Cuts plus different moving poles
Moving Cuts	?

14

The Multiperipheral Bootstrap

Now let us return to our simplest case, that of a fixed pole input. It will be recalled that we were able by appropriately choosing the input to generate a moving Pomeron pole as output, and to arrange this Pomeron to pass through any value we chose at t=0. We commented, however, that there was a difficulty, in that if the Pomeron existed, why wasn't it part of the input?

Is this bad? Why should we feel that the Pomeron should be something which can be repeatedly exchanged in a production process? Perhaps we should simply have stopped at some early step in our iteration, and said the Pomeron - (i.e. diffraction) is simply generated by some entirely different object, for example the fixed pole at $j=\alpha_0$ of our zeroth order input. Such a point of view requires us to believe that a Pomeron cannot itself be repeatedly exchanged in a multiperipheral sense, and it is hard to see why it cannot be. One of the motivations for the multiperipheral model, we recall, was the observation that at certain positive values of t_i, near physical masses, a production amplitude should be connected to a product of two-body amplitudes. But these amplitudes clearly contain Pomerons. Hence the production amplitude should contain repeated Pomerons.

Therefore, it appears that the most natural assumption is that the Pomeron itself - the output - must appear as part of the input and consequently there is a requirement of self-consistency between these input and output Pomerons. This self-consistency condition is summarized in the version of the multiperipheral model known as the multiperipheral bootstrap,[10)] which we discuss now. To begin with, let us take as input some sort of background together with the Pomeron itself, and let's assume that the Pomeron is simply a Regge pole. For simplicity, we may describe the background by some lower lying Regge pole or poles.

Let us think of the situation iteratively. In zero order, background (summarized for example by our fixed pole α_o) generates an approximate Pomeron. In first order, background plus the Pomeron generates the Pomeron. Thus in zeroth order, we had

$$T_2(t,j)T^{-1}(t,j) \propto j-\alpha_o(t) \tag{14.1}$$

as in eq. (13.8). In first order, we will have

$$T_2(t,j)T^{-1}(t,j) \propto j-\alpha_o(t)-\varepsilon(t,j) \qquad (14.2)$$

where the correction $\varepsilon(t,j)$ is due to adding the Pomeron itself to the input. Eq. (13.28) is an example. The function $\varepsilon(t,j)$ will contain, among other things, the AFS cut generated by the Pomeron itself. Thus we may write, near $j=1$,

$$T_2(t,j)T^{-1}(t,j) \propto j-\alpha_o(t)-\delta\alpha(t)-\gamma(t)\cdot$$

$$\cdot \ell n(j-\alpha_c(t)) \qquad (14.3)$$

where $\alpha_c(t) = 2\alpha_P(\frac{t}{4})-1 \qquad (14.4)$

and where the Pomeron $\alpha_P(t)$ is the leading pole of $T(t,j)$: Namely

$$\alpha_P(t) = \alpha_o(t)+\delta\alpha(t)+\gamma(t)\ell n(\alpha_P(t)-\alpha_c(t))$$

$$\equiv \alpha(t)+\gamma(t)\ell n(\alpha_P(t)-\alpha_c(t)). \qquad (14.5)$$

Thus in the absence of the AFS cut generated by the Pomeron itself, we would have $\alpha_P(t)=\alpha(t)$. With the cut, we have (14.5).

Suppose, for simplicity, that $\alpha(t)$ is a

straight line:

$$\alpha(t) = \alpha(o)+\alpha'(o)t. \tag{14.6}$$

Then near t=0 solutions of

$$\alpha_p(t) = \alpha(t)+\gamma(t)\ \ell n(\alpha_p(t)-\alpha_c(t)) \tag{14.7}$$

have the following properties:[11] The Pomeron intercept $\alpha_p(o)$ is below one, but above the intercept of the unperturbed trajectory $\alpha(t)$: We have $1>\alpha_p(o)>\alpha(o)$. We also note that $\alpha_p(o)$ lies above the intercept of the AFS cut, which occurs at $2\alpha_p(o)-1$. The Pomeron slope is less than the slope of the unperturbed trajectory, but is still positive: $0<\alpha_p'(o)<\alpha'(o)$. Finally, the Pomeron residue is small. Away from t near zero the situation is less easy to analyze, but the conjectured behavior is shown in fig. 14.1.

The fact that the Pomeron necessarily has intercept less than unity means that the cross section necessarily falls like a power of s. Recall that this same feature was also a consequence of attempting to impose self-consistency at t=0 between input and output poles in our simplest example, that with a fixed pole input. We also note that the Pomeron is strongly coupled for t>0 but weakly coupled for t<0. It thus has a some-

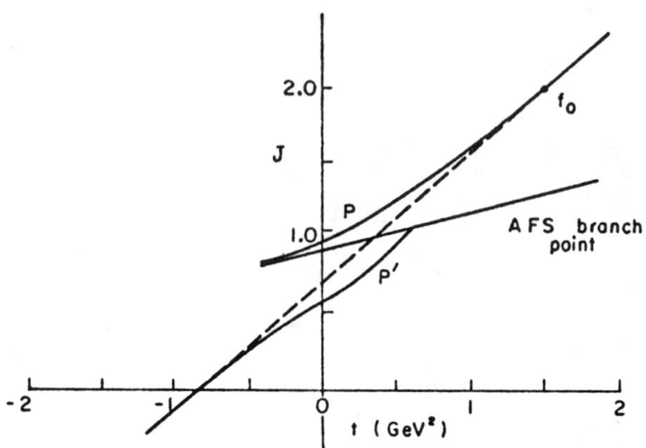

Fig. 14.1. Conjectured extrapolation away from t=0 of the P and (real part of) P' trajectories. The dashed line is the unperturbed trajectory, and the P' is a complex pair of poles on unphysical sheets. From ref. 11.

what dual nature according as $t>0$ or $t<0$; it is known as the "schizophrenic Pomeron."[11]

The schizophrenic Pomeron is on the physical j-plane sheet. But there are many sheets, because of the logarithmic j-plane AFS cut. And, there are also poles on these other sheets. Poles on the mth sheet are solutions of the equation

$$\alpha_m(t) = \alpha(t)+\gamma(t) \left[\ell n(\alpha_m(t)-\alpha_c(t))+im\pi\right]. \tag{14.8}$$

The solutions to this are evidently complex. Poles on the +mth and -mth sheet are at complex conjugate positions. As $t\to+\infty$ we have $\text{Re}\alpha_m(t)\to\alpha(t)$. $\text{Im}\alpha_m(t)$, in contrast, remains finite. As t becomes negative $\text{Re}\alpha_m(t)$ can freely pass below $\alpha_c(t)$, since

$\mathrm{Im}\alpha_m(t) \neq 0$, in contrast to the physical sheet pole. Furthermore, the residues of these poles are not particularly small for $t<0$.

The entire singularity structure of the multiperipheral output is, therefore, as follows: On the physical sheet is the Pomeron - steep and strong for $t>0$, flat and weak for $t<0$, with $\alpha_P(o)<1$. On unphysical sheets there are complex conjugate pairs of poles. These remain steep and strongly coupled throughout. Finally, there is also the AFS branch point. The whole mess is shown schematically in fig. 14.1, where the collection of unphysical sheet poles are approximated by a single (complex) Regge pole to which the name P' is given, in the hope that it has something to do with the physical P' trajectory.[11)]

The second iteration has led to a Pomeron - the schizophrenic Pomeron. It has also led to a cut, the AFS cut. The cut was not part of the input, and we are thus invited to proceed to a further iteration, in which this j-plane cut structure is also included in the input amplitude T_o. Needless to say, this becomes a very complex input, and the calculation of the predictions of the multiperipheral model with such an input have not been carried out. If it could be, it would doubtless yield an output even more complicated. We would still not have achieved a situation in which the output j-plane structure really coincides with the input in the multiperipheral picture.

Let us try to phrase the requirement of self-consistency more carefully. Suppose we take the particular multiperipheral model with

$$T_{2\to n}(p_1, p_2 \to q_1, \ldots, q_n)$$

$$= T_o(s_{12}, t_1) X(t_2) \ldots X(t_{n-2}) T_o(s_{n-1,n}, t_{n-1}) \tag{14.9}$$

as an example; recall that n can only be even in this model. We note that if we write (14.9) for n=2, we have

$$T_{2\to 2}(p_1, p_2 \to q_1, q_2) = T_o(s,t). \tag{14.10}$$

Thus, if (14.9) is to apply for all n, T_o should itself be the two-body scattering amplitude T(s,t). But then

$$T_{2\to n} = TXT \ldots XT \tag{14.11}$$

and using this in the s-channel unitarity relation yields the equations

$$T(t,j) = \frac{T_2(t,j)}{1 - \frac{1}{\pi^2} \frac{K(t)}{j+1} T_2(t,j)} \tag{14.12}$$

and

$$A_2(s,t) = \frac{1}{16\pi^2 s}\iint \frac{dt_1 dt_1'}{\sqrt{-\lambda}} T(s,t_1)T^*(s,t_1'). \tag{14.13}$$

Eqs.(14.12) and (14.13) constitute a highly nonlinear pair of homogeneous equations for the amplitude $A(s,t)$ (with Re T calculated from A by means of a fixed t dispersion relation). The solution of these (if it exists) is a self-consistent scattering amplitude, in that the output $T(s,t)$ exactly coincides with the input $T(s,t)$ occurring in (14.11). They constitute, for this model, the multiperipheral bootstrap equations.

It is obvious that this basic concept can be extended to more complicated multiperipheral models. For example, if off-mass shell behavior is included in (14.9), then equations (12.20) and (12.13) with $T_o(s,t;t_L,t_R)$ replaced by the actual two-body off-shell amplitude $T(s,t;t_L,t_R)$ form the bootstrap equations. Again, of course, they are a highly nonlinear set of equations.

We can also see that what we were trying feebly to do before was to solve this set of bootstrap equations by iteration, starting with the guess of a simple Regge pole behavior for the amplitude. And what we found, not surprisingly, was that a single Regge pole was in fact not self-consistent. No doubt the correct answer, if indeed a solution to the bootstrap equations exists at all, is exceedingly complex, and it goes without saying that no solution has yet been found.

Let us illustrate the possible use of the multiperipheral bootstrap idea with a contrived solution which in fact is not reasonable.[12)] Suppose the leading j-plane singularity is a flat cut at j=1:

$$T(t,j) \propto (j-1)^{\nu} \tag{14.14}$$

where ν is some constant. Then

$$A(s,t) \propto \frac{s}{(\ell n\ s)^{\nu+1}} \tag{14.15}$$

Now, assuming ReT(s,t)=0 or that it is negligible, we can calculate A_2: We get

$$A_2(s,t) \propto \frac{s}{(\ell n\ s)^{2\nu+2}} \tag{14.16}$$

and therefore

$$T_2(t,j) \propto (j-1)^{2\nu+1} + \text{constant.} \tag{14.17}$$

Now we use eq. (14.12), and here we make a crucial, and doubtless incorrect, step. Namely let us suppose that there is also some background, added to the flat cut as indicated by the constant in eq. (14.17). And let us suppose that this background exactly cancels the 1 in the denominator of eq. (14.12). Then, if $2\nu+1>0$, we get from (14.12) the result that

$$T(t,j) \propto (j-1)^{-2\nu-1}. \qquad (14.18)$$

Self-consistency between this result and (14.14) is thus possible if

$$\nu = -2\nu-1 \qquad (14.19)$$

and $$\nu = -\frac{1}{3}. \qquad (14.20)$$

Thus a flat cut like $(j-1)^{-1/3}$ is self-consistent.

The false step, as we have noted, is to assume that the background can cancel the 1 at all t. If it does not, at some t, then the 1 dominates $(j-1)^{2\nu+1}$ in the denominator and the argument fails. Nevertheless, this illustrates an attempt to find a self-consistent solution, and perhaps some day someone will find a real one.

15

Modified Multiperipheral Models

Let us try to summarize what we have done up to now. Our basic strategy has been to calculate elastic scattering by utilizing the s-channel unitarity relation

$$A(s,t) = A_{e\ell}(s,t)+A_I(s,t) \qquad (15.1)$$

where $$A_{e\ell}(s,t) = \frac{1}{16\pi^2 s}\iint \frac{dt_1 dt_2}{\sqrt{-\lambda}} T(s,t)T^*(s,t_2) \qquad (15.2)$$

is the two-body intermediate state contribution to A_2, and where

$$A_I(s,t) = \Sigma \int d\Phi_n \; T_{2\to n} \; T^*_{2'\to n} \qquad (15.3)$$

is the inelastic contribution to A. To do this we

require a model for the production amplitude $T_{2\to n}$, and this, motivated partially by experiments on production processes, partially by theoretical intuition, and mostly by calculational simplicity, we have attempted to provide through a multiperipheral model. That is, we have assumed that $T_{2\to n}$ has a factorized form and can be written as the product of a small number (one or two) of structures which, individually, depend on only a few of the variables involved in $T_{2\to n}$ itself.

At the first stage we guessed that the basic structure out of which $T_{2\to n}$ is formed was simply a fixed Regge pole, as in eq. (13.4), of which the field theoretical ladder in the ϕ^3 theory, as given in eq. (9.17), is a special case. With this input we calculated A_I, and found it to be a moving Regge pole, cf. eq. (13.10). Next we calculated A_I with a more complicated guess for the basic structure, namely a moving Regge pole, and finally we indicated what A_I might be with even more complicated inputs.

At this level, we were only calculating A_I from some given input in the multiperipheral model, and the models of $T_{2\to n}$ which we discussed were not meant to apply to the elastic amplitude $T_{2\to 2}=T$ itself. Implicit in this was the assumption that $A_{e\ell}$ in (15.1) is negligible compared to A_I, so that $A \simeq A_I$ and elastic scattering is predicted without further ado by the production amplitudes we have guessed. None of these models, we remind the reader, was notably successful experimentally.

At the next stage we complicated the kind of multiperipheral model we were dealing with; we assumed that the basic structure was itself the elastic amplitude which we wanted to find, as in eq. (14.10). Now our model for $T_{2\to n}$ was supposed to apply to n=2 as well; then the s-channel unitarity relation (15.1) becomes a self-consistent integral equation for the elastic amplitude, in that A_I is expressed in terms of T (and hence A), as, of course, is $A_{e\ell}$ through (15.2). We first attempted to solve this equation by guessing that the self-consistent solution to the equation would have as its leading singularities just a Regge pole; however, when we tried this guess in eq. (14.3) we found that it was not quite consistent. We found that a leading pole plus a background pole inserted as input produced a leading pole plus a cut on the left-hand side. We were, however, able to require self-consistency for the leading pole alone, and this partially self-consistent solution constituted the schizophrenic Pomeron model. The experimental success of the model was, however, minimal.

These two general classes of models - the one in which A_I is calculated through (15.3) from some input for $T_{2\to n}$, the other in which A is calculated self-consistently through (15.1) from a $T_{2\to n}$ which itself depends on T, or A, - summarize what has been done with multiperipheral models so far. What we should like to do in this chapter is to explore alterations of these two classes of models, in which the precise multiperipheral nature of the production amplitude may be lost. We shall see that permitting ourselves to modify the multiperipheral form for $T_{2\to n}$, often in ways which are either physically reasonably well motivated or which seem mathematically relatively minor, can lead to quite different predictions for high-energy scattering than those we have encountered so far.

The first elaboration we shall discuss is based on the first class of models summarized with above, those in which A_I is calculated from given input. In chapter 13 we essentially simply took $A=A_I$, and neglected A_2 in (15.1). Suppose we do not do this, but instead insert the calculated A_I in (15.1). Then, using (15.2), (15.1) becomes an integral equation for A. For our simplest input, in which $T_{2\to n}$ is the repeated exchange of a fixed Regge pole, A_I is a moving Regge pole. So let us take

$$A_I(s,t) = \beta(t)\ s^{\alpha(t)} \tag{15.4}$$

and, for simplicity, let $\beta(t)=\beta_o e^{bt}$ and $\alpha(t)=\alpha_o+\alpha' t$. Let us further assume that the elastic amplitude will be mostly imaginary for small t, and that we can set $T(s,t) \approx iA(s,t)$ in (15.3). Eq. (15.1) then becomes

$$A(s,t) = \beta_o\, s^{\alpha_o}\, e^{(b+\alpha'\ell n\, s)t}$$

$$+ \frac{1}{16\pi^2 s}\iint \frac{dt_1 dt_2}{\sqrt{-\lambda}}\, A(s,t_1)A(s,t_2). \qquad (15.5)$$

This is known as the AFS model. We note that the first iteration of this equation yields precisely the AFS cut discussed in chapter 10. This, we recall, was a logarithmic cut at $\alpha_c(t)=2\alpha(t/4) - 1$ which interfered constructively with the input pole A_I. We do not need to content ourselves with the first iteration, however, because eq. (15.5) can be solved exactly.[13] The solution, for $\alpha_o=1$, is

$$A(s,t) = 8\pi\beta_o \int_0^1 \frac{dy}{y}\left(1 - \sqrt{1 - \frac{\beta_o y}{4\pi(b+\alpha'\ell n\, s)}}\right)\cdot$$

$$\cdot J_o\left(2\sqrt{(b+\alpha'\ell n\, s)t\ \ell n\, y}\right). \qquad (15.6)$$

From this we see that at large t

$$A(s,t) \sim \frac{1}{t}\, e^{-2\sqrt{-t(b+\alpha'\ell n\, s)\ell n \frac{4\pi(b+\alpha'\ell n\, s)}{\beta_o}}} \qquad (15.7)$$

This is a t-dependence which drops off far too slowly to agree with experiment. It also turns out that $\sigma_{e\ell}(s)/\sigma_T(s)\to 1$ as $s\to\infty$, and the scattering, therefore, becomes purely elastic. Thus, in the solution of eq. (15.5), $A_{e\ell}$ dominates, and $A_{e\ell}>>A_I$, in contrast to the assumption that $A_I>>A_{e\ell}$ with which the model began. Experimentally, it would seem, the AFS model is a failure.

The same elaboration can, of course, be carried out, in principle, for other models of this class, which lead to more complicated functions $A_I(s,t)$. In each case, however, A_I is to be viewed as an input dictated by whatever multiperipheral model we use for $T_{2\to n}$. There is no feedback from the elastic scattering itself on A_I.

It is worth emphasizing, before leaving the AFS model, that the model does not correspond to any particular set of Feynman diagrams. For suppose the input was a propagator in ϕ^3 field theory. Then A_I is simply the ladder diagram of fig. 9.2. But then the solution to (15.5) corresponds only to the two-particle intermediate state contribution of the set of iterated ladder graphs, and as appeared in chapter 10, this solution has properties very different from the entire set of ladder graphs. Indeed, in field theory the plausible thing to do would be add all graphs of fig. 10.5. If this is done, the resulting elastic amplitude is given by an eikonalized ladder, as we have seen in chapter 10. We would find[14)]

$$T(s,t)=8\pi s\int_0^\infty bdb\ J_0(b\sqrt{-t})\left(\frac{e^{2i\chi(s,b)}-1}{2i}\right), \tag{15.8}$$

and for our simple input where A_I was given by (15.4),

$$\left(-\frac{1+e^{-i\pi\alpha(t)}}{\sin\pi\alpha(t)}\right)\beta(t)\ s^{\alpha(t)}$$

$$= 8\pi s\int_0^\infty bdb\ J_o(b\sqrt{-t})\chi(s,b). \qquad (15.9)$$

We have here included a signature factor on the left-hand side to emphasize that we use the entire ladder amplitude rather than just its absorptive part.

Choosing α and β as before, and approximating the signature factor by i yields

$$\chi(s,b) = \frac{i}{16\pi}\ \frac{\beta_o}{b+\alpha'\ln s}\ e^{-\frac{b^2}{4(b+\alpha'\ln s)}}. \qquad (15.10)$$

Thus, χ is imaginary; hence, the phase shift $\delta=0$ and $\chi=-i/2\ln\eta$. Hence (15.10) states that $\ln\eta$ is the Bessel transform of A_I:

$$-\frac{1}{2}\ln\eta = \frac{1}{8\pi s}\int_0^\infty \sqrt{-t}\ d\sqrt{-t}\ J_o(b\sqrt{-t})\ A_I(s,t). \qquad (15.11)$$

The model encompassed in (15.5), however, told us that A_I was the inelastic part of A. Thus (see appendix) from (15.5) we deduce

$$\frac{1-\eta^2}{4} = \frac{1}{8\pi s}\int_0^\infty \sqrt{-t}\; d\sqrt{-t}\; J_o(b\sqrt{-t})\; A_I(s,t). \tag{15.12}$$

The contrast between the AFS model and eikonalization is now clear: In one case we identify $1-\eta^2/4$ with a Regge pole; in the other we identify $-1/2\ \ell n\ \eta$ with a Regge pole. Note that when A_I is small, the expansions of these two expressions are:

AFS: $\eta=1-1/2\ X-1/8\ X^2+\ldots$

Eikonal: $\eta=1-1/2\ X+1/8\ X^2+\ldots$

where

$$X = \frac{1}{2\pi s}\int_0^\infty \sqrt{-t}\; d\sqrt{-t}\; J_o(b\sqrt{-t})\; A_I(s,t).$$

The cut term is the X^2 term: The opposite sign of the cut in the two cases, which we remind the reader, was discussed in chapter 10, is exhibited very clearly.

In sharp contrast to the point of view of the first class of models is the class of multiperipheral bootstrap models discussed in chapter 14, those in which $T_{2\to n}$ depended on T and in which (15.1) becomes a self-consistent homogeneous integral equation.

We shall discuss two modified versions of these models. In the first,[15)] we begin with the bootstrap equations we wrote down previously, namely, eqs. (14.12) and (14.13). Let us decompose $T_{e\ell}(t,j)$ into two parts; a singular part $T_s(t,j)$ which contains the leading j-plane singularity near j=1 for small t, and a nonsingular part $T_{ns}(t,j)$ which is analytic in j near j=1 when t is small. Evidently the asymptotic behavior in s is controlled by T_s; T_s may thought of as the "Pomeron" contribution. In contrast, T_{ns} represents a kind of "background" such as would be provided by the exchange of lower lying Regge trajectories.

Eq. (14.12) can then be written

$$T(t,j)=\frac{T_s(t,j)+T_{ns}(t,j)}{1-\frac{1}{\pi^2}K_s(t,j)T_s(t,j)-\frac{1}{\pi^2}K_{ns}(t,j)T_{ns}(t,j)}, \tag{15.13}$$

where K_s and K_{ns} are kernels associated with the "Pomeron" and "background" respectively. Both are expected to be analytic near j=1 and t=0.

The goal is to study (15.13) near the leading singularity in j for small t. Let us, therefore, expand the nonsingular functions T_{ns}, K_s, and K_{ns} in powers of j=1 and t. Then (15.13) takes the form

$$T(t,j) = \frac{T_s(t,j)+a_o+a_1(j-1)+a_2t+\ldots}{(b_o+b_1(j-1)+b_2t+..)-(c_o+c_1(j-1)+c_2t+..)T_s(t,j)}. \tag{15.14}$$

The various constants in (15.14) are in principle determined by the "background" and by the (presumably) known kernels; they must, however, be such as to permit a self-consistent solution to (15.14) and (14.13). It turns out that a solution does exist if b_o is zero and if c_o and c_1 are zero, too. In this event (15.14) becomes

$$T(t,j) = \frac{\beta T_s(t,j) + \beta'}{j-1-\gamma t\ T_s(t,j)}\ , \tag{15.15}$$

with an appropriate redefinition of constants, and neglecting terms which are not important in the $j\to 1$, $t\to 0$ limit.

It should at this point be made clear that c_o and c_1 cannot be zero in the simplest kind of multiperipheral models we have discussed up to now. For as can be seen from (12.23), a kernel like K_s is positive definite at $t=0$. Thus, some unusual off-mass shell dependence of the elastic amplitude is required to permit (15.15) to be valid; it is in this sense that the model is a modified multiperipheral one.

Eqs. (15.15) and (14.13) can be solved exactly if, as usual, we assume $\mathrm{Re}T(s,t) \ll A(s,t)$. The solution is

$$T(t,j) = \frac{1}{\gamma t}\left(\frac{j-1}{\sqrt{(j-1)^2 - R_o{}^2 t}} - 1\right), \tag{15.16}$$

which is an amplitude with two branch points at $j=1\pm iR_o\sqrt{-t}$ reducing to a double pole at $t=0$.

In the s,t language this translates into

$$T(s,t) = is\left(\frac{8\pi R_o^{\,2}}{\beta}\right)\left(\frac{J_1(R_o\sqrt{-t}\ \ell n\ s)}{R_o\sqrt{-t}}\right). \quad (15.17)$$

This amplitude can also be expressed in impact parameter language, and we find

$$T(s,b) = \frac{8\pi is}{\beta\ \ell n\ s}\ \ \Theta(R_o\ \ell n\ s - b);$$

the scattering particle thus looks like a graying disk with a growing radius.

There are a number of other predictions, as listed below.[16)]

(i) The total cross section grows logarithmically with s: We have

$$\sigma_T \to \left(\frac{4\pi R_o^{\,2}}{\beta}\right)\ell n\ s + \beta'. \quad (15.18)$$

(ii) The elastic cross section becomes a constant:

$$\sigma_{e\ell} \to \frac{4\pi R_o^{\,2}}{\beta^2}\ . \quad (15.19)$$

(iii) The elastic differential cross section is

$$\frac{d\sigma}{dt} \Big/ \left(\frac{d\sigma}{dt}\right)_{t=0} = \left(\frac{2\, J_1(R_o\sqrt{-t}\ \ell n\ s)}{R_o\sqrt{-t}\ \ell n\ s} \right)^2 . \tag{15.20}$$

Thus, there is $(\ell n\ s)^2$ shrinkage and there are dips moving toward $t=0$ like $(\ell n\ s)^{-2}$.

(iv) The partial cross sections for producing $2n+2$ particles are

$$\sigma_n = \frac{4\pi R_o^2}{\beta^2} \frac{\gamma(n,\beta\ell n\ s)}{(n-1)!} + \beta' s^{2\alpha-2} \frac{(\beta\ell n\ s)^n}{n!} \tag{15.21}$$

where γ is an incomplete gamma function and where $\alpha=1-\beta$. Thus, there is a "diffractive" component, the first term, which yields a contribution to σ_n which is asymptotically constant for $n \lesssim \beta\ \ell n\ s$ and zero beyond. There is also a multiperipheral component, the second term, in (15.21), which vanishes with a power of s; yet when the multiperipheral parts are added together they contribute a constant to σ_T. Diffractive dissociation is then responsible for the $\ell n\ s$ rise in σ_T.

(v) The multiplicity is logarithmic, and the correlation coefficients are $f_n \sim (\ell n\ s)^n$; there exist long-range correlations.

The final modified model we wish to discuss is one in which we alter the multiperipheral structure of $T_{2\to n}$.[17)] A fundamental principle which is not taken into account in the multiperipheral model is that of full (n→m) s-channel unitarity. Only 2→2 unitarity is imposed, in virtue of eq. (15.1). A simple way of imposing 2→n unitarity, at least, is to introduce absorption. The assumption of factorizability does not allow for initial or final state interactions; in a fully unitary theory these should be expected to exist.

Let us suppose, then, that our model of production is

$$T_{2\to n} = \sqrt{S}\; T^{MP}_{2\to n}, \qquad n>2, \tag{15.22}$$

where S is the elastic S-matrix and $T^{MP}_{2\to n}$ stands for the production amplitude of the multiperipheral bootstrap model. From (15.22) we see that in the impact parameter representation

$$A(s,b) = A_{e\ell}(s,b) + A_I(s,b) \tag{15.23}$$

where

$$A_{e\ell}(s,b) = \frac{1}{8\pi s}\,|T(s,b)|^2 \tag{15.24}$$

is the two-body intermediate state contribution and where the inelastic contribution is

$$A_I(s,b) = S(s,b)\, A_I^{MP}(s,b). \tag{15.25}$$

Now, if ReT is as usual assumed to be less than A, we have

$$S(s,b) = \eta(s,b) \tag{15.26}$$

while

$$A(s,b) = 4\pi s(1-\eta(s,b)). \tag{15.27}$$

Thus, (15.25) becomes

$$A(s,b) = 4\pi s + A_I^{MP}(s,b) - \sqrt{(4\pi s)^2 + (A_I^{MP}(s,b))^2}. \tag{15.28}$$

Now our self-consistent problem is defined by (15.28) together with the usual expression for A_I^{MP} from the multiperipheral bootstrap, viz.):

$$T_I^{MP}(t,j) = \frac{T_2(t,j)}{1-\frac{1}{\pi^2}K(t,j)T_2(t,j)} - T_2(t,j), \tag{15.29}$$

as in (14.12) as well as the normal definition of A_2 in (14.13). The exact solution to this problem can be written down analytically, and is

$$T(s,t) = 4\pi i s \frac{R_o \ln s}{\sqrt{-t}} J_1(R_o\sqrt{-t} \ln s). \tag{15.30}$$

We may again rewrite this in impact parameters

language as

$$T(s,b) = 4\pi is\ \Theta(R_o\ \ell n\ s - b);$$

the scattering is, therefore, from a black disk of radius R ℓn s. In the j-plane we have

$$T(t,j) \propto ((j-1)^2 - R_o{}^2 t)^{-3/2}$$

and the singularity structure is two complex conjugate branch points of $(j-\alpha_c)^{-3/2}$ character at $\alpha_c = 1 \pm iR_o\sqrt{-t}$, which coalesce into a third order pole at t=0. This is precisely saturation of the Froissart bound, the same conclusion that resulted from QED in field theory. There is a total cross section growing like $(\ell n\ s)^2$, the ratio $\sigma_{e\ell}/\sigma_T$ approaches 1/2, there is $(\ell n\ s)^2$ shrinkage and there are dips in $d\sigma/dt$ moving into t=0 like $(\ell n\ s)^2$.

Further consequences[18)] are that the multiplicity is logarithmic, that scaling in inclusive distributions fails, but only by an overall logarithmic factor, and correlations are long range. It is interesting to note that some of these consequences for production differ from those of the QED model.

To conclude, these elaborated and modified multiperipheral models hopefully provide a suggestion of the kind of freedom and kinds of conclusions which are available and can be achieved if one is only willing to broaden one's outlook. It is important to keep in mind that the narrow multiperipheral models are justified neither by experiment nor by theory, and there is every reason to look further.

REFERENCES

1. D. Amati, S. Fubini and A. Stanghellini, Nuovo Cim. 26, 896 (1962).
2. G. Chew and A. Pignotti, Phys. Rev. Letters 20, 1078 (1968), Phys. Rev. 176, 2112 (1968).
3. G. Chew, F. Low and M. Goldberger, Phys. Rev. Letters 22, 208 (1969).
4. C.I. Tan, Phys. Rev. D3, 790 (1971).
5. S. Pinsky and W. Weisberger, Phys. Rev. D2, 1640, 2365 (1970).
6. C. De Tar, Phys. Rev. D3, 128 (1971).
7. D. Silverman and C.I. Tan, Phys. Rev. D2, 233 (1971).
8. N. Bali, A. Pignotti and D. Steele, Phys. Rev. D3, 1167 (1971).
   D. Silverman and C.I. Tan, Nuovo Cim. 2A, 489 (1971).
9. J. Ball and G. Marchesini, Phys. Rev. 188, 2209, 2508 (1969).
   G. Chew, T. Rogers and D.R. Snider, Phys. Rev. D2, 765 (1970).
10. G.F. Chew and D.R. Snider, Phys. Rev. D1, 3453 (1970).
11. G.F. Chew and D.R. Snider, Phys. Rev. D3, 420 (1971).
12. D. Branson, Nuovo Cim. 3A, 271 (1971).
13. D. Amati, M. Cini and A. Stanghellini, Nuovo Cim. 30,193 (1963).
14. S.C. Frautschi and B. Margolis, Nuovo Cim. 56, 1155 (1968).
15. J. Ball and F. Zachariasen, Physics Letters 40B, 411 (1972).

16. J. Ball and F. Zachariasen, Physics Letters 41B, 525 (1972).
17. J. Finkelstein and F. Zachariasen, Physics Letters 34B, 631 (1971).
18. L. Caneschi and A. Schwimmer, Nucl. Phys. B44, 31 (1972).

PART V

OTHER MODELS

Each field-theoretic or multiperipheral model provides us with a specific form for the n-particle production matrix element. Such models can be regarded as "complete theories" in the sense that, in principle at least, they are capable of describing every detail of high energy processes. In particular, each of them is capable of a specific statement about the character of the Pomeron, and indeed this question has occupied center stage so far.

We wish, now, to turn to a set of models much less ambitious, and, correspondingly, much less precisely formulated. These are models which focus on only one or another aspect of high energy reactions and which do not, in general, purport to be able to describe all features of these reactions. They might, perhaps, be called semiphenomenological models, and in most of them the nature of the Pomeron is left unspecified as not particularly germane to the limited set of applications which can be made.

We shall discuss droplet models, statistical models, diffractive models and hybrid models. Each of them starts with a guess, motivated mostly by intuition, at some feature of high energy processes and attempts to extrapolate from this to some slightly wider domain of phenomena. Thus in the droplet, one guesses that elastic scattering is like classical physics. The Pomeron is thereby assumed to be a fixed pole at $j=1$, and little can be said about production processes. In statistical models one assumes various statistical or thermodynamic descriptions are possible for the set of particles produced in a high energy reaction. These models have little or nothing to say about diffraction, or about the Pomeron. In the diffractive model, one pretends that all particle production is just like diffraction dissociation; hence, the Pomeron is usually implicitly assumed to be just a Regge pole, but the model can really work with any kind of Pomeron, and it will take what it gets. And, obviously, it has nothing whatever to say about elastic scattering.

16

The Droplet Model

The impact parameter representation allows us to write the elastic amplitude in a form which is suggestive of a well identified structure in configuration space. Let us start with the representation of eq. (I.19).

$$T(s,t) = 8\pi s \int_0^\infty bdb\ J_o(b\sqrt{-t})\ \frac{\eta e^{2i\delta} - 1}{2i} \quad (16.1)$$

which expresses the two-body scattering amplitude as a Bessel transform of the phase shift and absorption coefficient. In the droplet model,[1] which we are going to discuss now, one starts with the assumption that elastic scattering is due to absorption only, i.e. $\delta=0$. In this case T=iA is purely imaginary and

$$A(s,t) = 4\pi s \int_0^\infty bdb\ J_o(b\sqrt{-t})\ (1-\eta(s,b)). \quad (16.2)$$

If we continue one step further and maintain that η is actually independent of s we arrive at a classical picture in the sense that we attribute to the particles a sort of rigid structure which is independent of the principal dynamical variable - the scattering energy. This results in the form

$$T = isf(t) \tag{16.3}$$

which corresponds to a fixed j-plane pole and which is often discussed in the literature in spite of the fact that it leads to theoretical difficulties (see the discussion in appendix D). Eq. (16.2) can now be rewritten as

$$f(t) = 2\int d^2b \; (1-\eta(b)) \; e^{i\vec{q}\cdot\vec{b}} \tag{16.4}$$

where we have used an explicit two-dimentional form in which $\vec{b}$ is the impact parameter and $\vec{q}^{\,2}=-t$.

If we consider the two scattering particles as blobs of absorbing matter of some given spatial structure then the amount of absorption will depend on the overlap of these two blobs. In the high-energy limit we think of the scattering as an instantaneous reaction between two contracted objects that are described by a density distribution $D(\vec{b})$ in the transverse plane. One can then expect

$$\chi = i\frac{K}{2}\int d^2b' \; D_A(\vec{b}') \; D_B(\vec{b}'-\vec{b}) \qquad \eta=e^{2i\chi} \tag{16.5}$$

where K is the absorption coefficient. This represents the exponential attenuation of a wave passing through an absorbing medium. For small absorption this reduces to the linear approximation

$$\eta(b) = 1-K\int d^2b' D_A(\vec{b}')\ D_B(\vec{b}'-\vec{b}) \tag{16.6}$$

which means that

$$f(t) = 2K\int d^2b\ e^{i\vec{q}\cdot\vec{b}} \int d^2b' D_A(\vec{b}')D_B(\vec{b}'-\vec{b}). \tag{16.7}$$

In the case of the scattering of identical particles $AA\to AA$, and presumably also $A\bar{A}\to A\bar{A}$ and $A_iA_j\to A_iA_j$ where A_i and A_j are in the same isospin multiplet,

$$f(t) = 2K\ |D_A(t)|^2 \tag{16.8}$$

where one uses the obvious definition

$$D(t) = \int d^2b\ e^{i\vec{q}\cdot\vec{b}} D(\vec{b}). \tag{16.9}$$

At the point t=0 one can use eq. (16.7) to deduce the total cross section,

$$\sigma_T^{AB} = f(o) = 2K\int d^2b D_A(\vec{b})\int d^2b D_B(\vec{b}) \tag{16.10}$$

which has a factorized form and implies that the total cross section is proportional to the "amount of matter" in each of the two particles.

Since the nature of the charge distribution inside the nucleons is known from the electromagnetic form factors, it was first suggested that one should use the same form factors to describe the distribution of total hadronic matter. One then identifies D(t) in eq. (16.9) with the electromagnetic form factor of the proton and expects to find the following structure for pp scattering:

$$\frac{d\sigma}{dt} = \left(\frac{d\sigma}{dt}\right)_{t=o} G^4(t). \tag{16.11}$$

We saw in chapter 4, in figs. 4.1 and 4.3, that the high-energy pp data fall far below this value One may, however, still take the point of view that the approximation involved in deducing (16.6) from (16.5) is to blame for this discrepancy; in other words, K is not small enough to justify expanding $e^{2i\chi}$. If, indeed, one does not use the linear approximation but calculates the exact f(t) from (16.4) and (16.5) one may obtain an exponential fall-off similar to that required by the data. In addition, the full eikonal form predicts a diffraction pattern,[2)] i.e., zeroes in $\frac{d\sigma}{dt}$. The result of such a calculation can be seen in curve A in fig. 4.1. Note the nice agreement with the shape of the curve and the location of the dip.

If one allows the coefficient K to become complex, and therefore $\delta \neq 0$, the sharp dip gets covered up. This is shown in curve B of fig. 4.1

which includes a real part allowed by the data. Within this wide range one can clearly obtain a good fit. One should also note that once the linear approximation is given up, eq. (16.10) no longer holds, and factorization of total cross sections can be at best only an approximate property.

The droplet model has, of course, very different properties from a multiperipheral model. In the latter one expects the average impact parameter, or the radius of interaction, to grow logarithmically with s. The reason for this is that at higher energies longer chains of exchanges become important. A multiperipheral chain with n rungs, based on Feynman diagrams with exchanges of particles of average mass μ, leads to a diffraction radius given by $\langle b^2\rangle \propto \frac{n}{\mu^2} \propto \ell n\ s$. This follows from the fact that each propagator corresponds in the transverse configuration space to an average separation of $\frac{1}{\mu}$ and the various propagators combine in a random walk fashion.[3)]

The Regge pole exchange in the multiperipheral model provides us with a combined s and t dependence. In addition, there is further unspecified t dependence in the Regge residue functions. One may associate this t-dependence with an internal structure of the colliding particles which can be described by a droplet model. Thus one may replace the point interactions between the two droplets by a multiperipheral exchange[4)] leading to

$$\chi = \frac{i}{2}\int d^2b_1 D_A(\vec{b}_1)d^2b_2D_B(\vec{b}_2)g(\vec{b}_1-\vec{b}_2-\vec{b}). \tag{16.12}$$

In eq. (16.5) $g(\vec{a})$ was given by $K\delta^{(2)}(\vec{a})$ because the original droplet model calls for pointlike interactions between the two blobs. A characteristic multiperipheral correction would be of the form

$$g(\vec{a}) \propto e^{-\frac{\vec{a}^2}{\alpha+\beta\ell n s}} \tag{16.13}$$

corresponding to an interaction radius which is growing logarithmically with energy. It will still be true that in the linear approximation the total cross section factorizes

$$\sigma_T^{AB} \approx 2\int d^2bg(\vec{b})\int d^2bD_A(\vec{b})\int d^2bD_B(\vec{b}) \tag{16.14}$$

and the higher order terms in the eikonal expansion will now correspond to cuts or absorptive corrections. This adds still more parameters to the model and it remains to be seen to what extent the forthcoming very high-energy data will lend itself to such fits. The advantage of (16.12) over a simple multiperipheral picture is that it provides a model for the t dependence, a feature that is left almost free in any normal Regge-type theory, whose predictions concern mainly the energy dependence of cross sections.

Let us return now to equation (16.3) and discuss an alternative way of obtaining information about the structure function f(t). We note that the simple form A=sf(t) when incorporated with the s-channel unitarity formula

$$A(s,t) = A_I(s,t) + \frac{1}{16\pi^2 s}\iint \frac{dt_1 dt_2 \theta(-\lambda)}{\sqrt{-\lambda(t,t_1,t_2)}} \cdot$$

$$\cdot\ T(s,t_1)\ T^*(s,t_2) \qquad (16.15)$$

leads to an integral equation for f(t):

$$f(t) = \frac{1}{s} A_I(s,t) + \frac{1}{16\pi^2}\iint \frac{dt_1 dt_2 \theta(-\lambda)}{\sqrt{-\lambda(t,t_1,t_2)}} \cdot$$

$$\cdot\ f(t_1)\ f(t_2). \qquad (16.16)$$

$A_I(s,t)$ is the contribution of all inelastic channels to the unitarity equation. For this equation to be self-consistent we must, of course, have

$$A_I(s,t) = sf_o(t) \qquad (16.17)$$

and for any value of $f_o(t)$ the integral equation should give the corresponding f(t) and vice versa.

If we start from a form $f(t) \propto e^{bt}$, a reasonable representation of the data, then we can calculate the elastic part of the unitarity equation through

$$\iint \frac{dt_1 dt_2}{\sqrt{-\lambda(t,t_1,t_2)}} e^{bt_1} e^{bt_2} = \frac{\pi}{2b} e^{bt/2} \qquad (16.18)$$

and find that it has half the slope of the elastic amplitude. $f_o(t)$ will then be of the form

$$f_o(t) = f(o)\, e^{bt} - \frac{1}{32\pi b}(f(o))^2\, e^{bt/2}. \qquad (16.19)$$

From fig. 4.1 we obtain for pp scattering

$$\frac{d\sigma}{dt}(t=o) = \frac{1}{16\pi}\,(f(o))^2 \approx 70\,\frac{mb}{GeV^2} \qquad (16.20)$$

and $b \simeq 6\ GeV^{-2}$; hence, it follows that in this case

$$f_o(t) \approx 37\, e^{6t} - 6\, e^{3t}\ mb. \qquad (16.21)$$

It is clear that this quantity passes through zero around $t \simeq -0.6\ GeV^2$.

$A_I(s,t)$ was first investigated by Van Hove,[5] who named it the overlap integral, and suggested on the basis of highly simplified production models that it might have a pure exponential behavior. In fact, however, we see that if the scattering amplitude is purely imaginary (as implicitly assumed in the derivation above) the overlap integral has an exponential behavior for $-t \lesssim 0.3\ GeV^2$, but for larger -t it decreases faster and changes its sign around $t=-0.6\ GeV^2$. It is easy to show that even if the amplitude has a non-zero phase this conclusion does not change. Thus if

$$f(t) = f(o)e^{bt+ict} \qquad T=isf(t) \tag{16.22}$$

we will find

$$f_o(t) = f(o)e^{bt}\cos ct - \frac{1}{32\pi b}(f(o))^2 e^{t\frac{b^2+c^2}{2b}} \tag{16.23}$$

which is not too different from (16.19) as long as $c<b$. In a Regge model one would write

$$\begin{aligned} T &= i\beta(t)e^{\alpha' t \ln s}(1+i\cot\frac{\pi\alpha}{2}) \\ &= -\beta(t)\, e^{\alpha' t \ln s}\, \frac{e^{-i\frac{\pi\alpha' t}{2}}}{\sin\frac{\pi\alpha' t}{2}} \end{aligned} \tag{16.24}$$

which is similar to the form (16.22) and implies a value of c of the order of 1 GeV^{-2}. This has to be compared with $b \simeq 6\ \text{GeV}^{-2}$, which means that the main conclusion about the expected zero in A_I still holds.

The droplet model puts the emphasis on the simple structure of the colliding particles rather than on that of the secondary distributions. The mathematical formulation of eq. (16.5) gives a definite meaning to elastic scattering in this model. The same kind of thinking can be continued into the realm of inclusive distributions. This type of model was the basis for the idea of limiting fragmentation (see chapter 2). The limiting distributions, when measured in the target frame, are assumed to be independent of the projectile and reflect the structure of the target. The qualitative consistency of this picture is reflected in the fact that the q_T distribution of inclusive spectra - which is then determined by $D(\vec{b})$ - is much less peaked than that of elastic scattering. The latter is given in (16.5) by a convolution of two matter distributions and should therefore be wider in transverse configuration space and narrower in q_T.

In dealing with inclusive distributions we add another degree of freedom. Whereas in eq. (16.5) we were interested only in the transverse structure we investigate in inclusive distributions the longitudinal x structure as well. No x dependence was discussed in the model for elastic scattering. Such additional degrees of freedom can be always added by writing

$$D(\vec{b}) = \sum_i \int_0^\infty f_i(x,\vec{b})\, dx \qquad (16.25)$$

where i can designate the type of constituent in the particle whose matter density is described by $D(\vec{b})$. In addition, the various types of constituents may have different sorts of interactions thus leading to more complicated structures than the simple model described above.

The additional degrees of freedom mentioned here can be understood in two basically different ways. The first is to associate them with the produced particles themselves, as for instance by choosing

$$f_i(x,\vec{b}) = \frac{1}{\sigma_{AB}} \int e^{i\vec{b}\cdot\vec{q}} \rho^{i}_{AB}(x,\vec{q}_T)d^2q_T \quad (16.26)$$

to describe the fragmentation distributions of either particle A or B (at positive or negative x respectively). By this choice we are assured of a probability interpretation of $f_i(x,\vec{b})$, in view of eq. (2.6). It is, of course, in accord with the interpretation of fragmentation given above and leads to a rough agreement with the observed t dependence. However, it is not really motivated by any known theoretical model.

The alternative is to discuss a constituent model such as the parton-quark model, in which $f_i(x,\vec{b})$ denotes the distributions of the six types of quarks and anti-quarks in the system. The usual applications of such models are in deep inelastic electron scattering.[6] In that case, one is interested in $\int f_i(x,\vec{b})d^2b$, since only the x-dependence is measured. Thus we see that the information from deep inelastic electron scattering is, in a sense, orthogonal to that discussed throughout this chapter. In fact, the application of the parton model to hadron-hadron scattering and inclusive production calls for a host of new assumptions. The beauty of deep inelastic scattering is in the point interaction of the photon vertex. The hadron scattering problem is a collective effect of constituent interactions and the pointlike parton character may exhibit itself, if at all, only in rare cases such as very high q_T inclusive production, as mentioned in chapter 2.

17

Statistical Models

We turn first to the uncorrelated jet model for multiparticle production.[5,7] It is based on two simple features: First, it imposes on the produced particles the requirement of limited transverse momenta, leaving them essentially with only a longitudinal degree of freedom. This results in the formation of jets of secondaries. The second simplifying assumption is that the various emitted particles are uncorrelated. This idea implies that the prong distributions can be approximated by Poisson distributions (see the discussion in chapter 1). Although this is no longer true experimentally at higher energies, one can conceive of a more complicated model that will incorporate in it some of the basic features of the uncorrelated jet model. We discuss, therefore, some of the characteristics of this approach.

Let us start the discussion with a look at phase space formulae.[8] The phase space of an n-particle configuration can be evaluated in the c.m. frame by

$$F_n(\sqrt{s}) = \int \prod_i \frac{d^3q_i}{\omega_i} \delta^{(3)}(\sum_i \vec{q}_i)\delta(\sum_i \omega_i - \sqrt{s})$$

$$= s^{n-2} \Psi_n\left(\frac{\sqrt{s}}{\mu}\right) \qquad (17.1)$$

where we assume for simplicity that all particles have the same mass μ. This coincides with the definitions in the Introduction up to inessential numerical factors. The function $\Psi_n(\frac{\sqrt{s}}{\mu})$ is regular in the limit $\mu \to 0$ or $\sqrt{s} \to \infty$ and therefore $F_n(\sqrt{s})$ grows asymptotically like s^{n-2}. If, however, we deal with a problem that has only one dimension we find

$$f_n(\sqrt{s}) = \int \prod_i \frac{dq_i}{\omega_i} \delta(\sum_i q_i)\delta(\sum_i \omega_i - \sqrt{s}) = \frac{1}{s}\psi_n\left(\frac{\sqrt{s}}{m_T}\right) \qquad (17.2)$$

where we have now designated the common mass by m_T. One can refer to $f_n(\sqrt{s})$ as a longitudinal phase space since all transverse dependence is wiped out and the mass is replaced by the appropriate transverse mass. We expect that if one were to use a formula like eq. (1.1) or (2.1) and replace the matrix element by a strongly damped exponential in q_T the resulting expression would have the characteristics of longitudinal phase space (17.2).

In order to perform the integral in (17.2)

it is convenient to switch to rapidity variables. It is then easy to show that

$$f_2(\sqrt{s}) = \frac{2}{\sqrt{s}\ \sqrt{s-4m_T^2}}$$

$$f_n(\sqrt{s}) = \int dy_1 \ldots dy_{n-2}\ f_2(\sqrt{s'}) \qquad (17.3)$$

where $s'^2 = s-2\sqrt{s}E+M^2 \qquad E = \sum_{i=1}^{n-2} E_i$

$$Q = \sum_{i=1}^{n-2} q_i \qquad M^2 = E^2 - Q^2. \qquad (17.4)$$

The asymptotic features are determined by the overall rapidity volume which leads to

$$\psi_n\left(\frac{\sqrt{s}}{m_T}\right) \to \left(\ln \frac{\sqrt{s}}{m_T}\right)^{n-2} \qquad n \text{ fixed, } s\to\infty. \quad (17.5)$$

This means that the asymptotic form of each $f_n(\sqrt{s})$ is essentially s^{-1}, in sharp contrast to the growing trend like s^{n-2} of $F_n(\sqrt{s})$. It is interesting to note that the energy variation of $f_n(\sqrt{s})$ has features similar to experimental exclusive cross sections as shown in fig. 17.1.

The one-particle distribution in the n-particle configuration of the longitudinal phase space is simply $f_n(W)$ where W is the remaining missing mass,

$$W^2 = s-2\sqrt{s}\ \omega+m_T^2. \qquad (17.6)$$

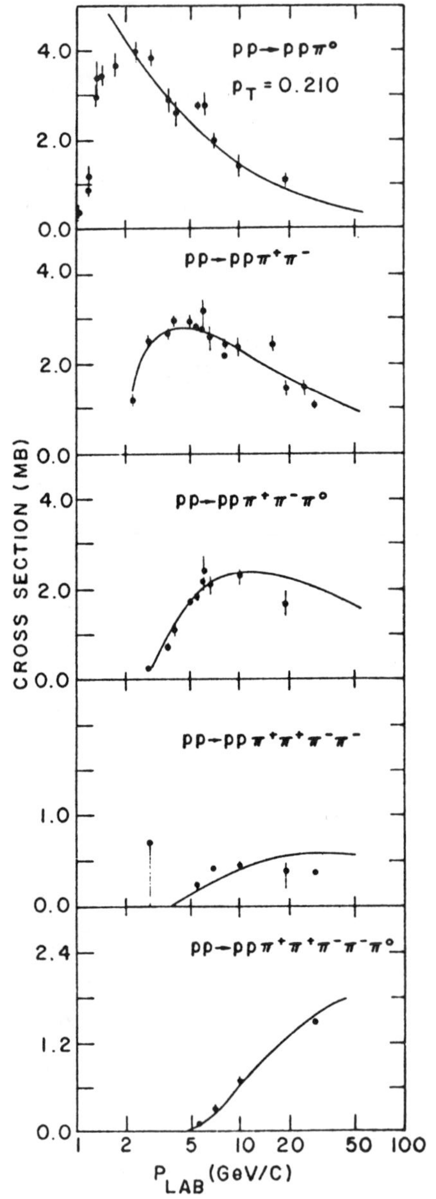

Fig. 17.1. Compilation of cross sections for the processes $pp \to pp + n\pi$ where $n=1,\ldots,5$. The shown curves are predictions of LPS with $p_T = 0.210$ GeV/c with an arbitrary normalization. Taken from ref. 9.

Here ω is the energy of the particle in question in the c.m. system. Since $f_n(W)$ is a decreasing function of W at high W and fixed n one finds that there is a tendency towards production of particles at higher ω values.[10] This effect comes about because of energy-momentum conservation and the limited q_T range.

The second element of the uncorrelated jet model is the statistical approach.[11] One may start by assuming that the n-particle configuration has a Poisson distribution $\frac{z^n}{n!}$ where z is essentially the average number of particles produced. Once one goes into a more detailed situation,[12] like $p+p \to N+N+n\pi$, one may further ask how the n pions divide up into the various charge states. A simple approach to this question is once again a statistical one - assume that all the ways of forming the required total isospin are of equal weight. Then one may define a probability distribution

$$W(n_+,n_-,n_o) = \left[\sum_\alpha |\langle n_p,n_n,n_+,n_-,n_o|I,I_3,\alpha\rangle|^2\right] \cdot \left[\sum_{\alpha\{n\}} |\langle\{n\}|I,I_3,\alpha\rangle|^2\right]^{-1} \quad (17.7)$$

where α designates the additional quantum numbers required to specify a state with a given I, I_3. Here one assumes that

$$n_+ + n_- + n_o = n \qquad n_p = 2 - n_+ + n_-$$

$$n_n = n_+ - n_- = 2 - n_p$$

and $W(n_+n_-n_o)$ will therefore describe the probability of finding a specific $n_+n_-n_o$ state for a given n. It has been shown by Cerulus[13)] that

$$W(n_+,n_-,n_o) = \frac{n!}{n_+!n_-!n_o!} \frac{2!}{n_p!n_n!} \cdot$$

$$\cdot\ 3\cdot 2^{-(n_+ + n_- + 3)} \int_{-1}^{1} dx\,(1+x)^{n_+ + n_- + 2}\ x^{n_o}. \tag{17.8}$$

The prominent feature here is the binomial distribution, which shows us that the statistical assumption leads again to a Poisson distribution, this time in individual pions, on which one has simply to superimpose charge conservation:

$$\frac{z^n}{n!} \frac{n!}{n_+!n_-!n_o!} \delta(n_+ - n_- - q)$$

$$= \frac{z^{2n_- + q}}{n_-!(n_- + q)!} \frac{z^{n_o}}{n_o!} . \tag{17.9}$$

Thus we have a Poisson distribution for π^o multiplied by a distribution for the ensemble of charged pions with overall charge $q = n_n \leq 2$. This turns out to be generated by a Bessel function[14)]

$$I_q(2x) = \sum_n \frac{x^{q+2n}}{n!(n+q)!} \xrightarrow[n\to\infty]{} \frac{(2x)^{2n}}{(2n)!\sqrt{\pi n}} \tag{17.10}$$

which differs slighly from a Poisson distribution in charged pairs for high values of n.

Using the longitudinal structure and the statistical approach one can try to write down an explicit form for the n-particle production cross section:

$$\sigma_n \propto \int \prod_{i=1}^{n} \frac{d^3q_i}{\omega_i} \frac{dp_1}{E_1} \frac{dp_2}{E_2} \frac{z^n}{n!} W(n_+ n_- n_o) \cdot$$
$$\cdot F_N(p_1) F_N(p_2) \prod_{i=1}^{n} G_\pi(q_i); \qquad (17.11)$$

this is once again written for the case of $p+p \to N_1+N_2+n$ pions.

We quote this particular example since it has recently been discussed in detail in the literature.[12,15] The functions F_N and G_π include a transverse momentum cutoff and some x dependence as well. The choice made in ref. 15 is, essentially,

$$F_N(p) = \exp(-R^2 p_T^{\,2})$$

$$G_\pi(q) = \exp(-\lambda|x| - R^2 q_T^{\,2}). \qquad (17.12)$$

The parameters are chosen as $R^2 = 6\ \text{GeV}^{-2}$ and $\lambda = 5$. The average transverse momentum is then approximately constant and in agreement with the data (fig. 17.2). The parameter $z = 3\ \text{GeV}^{-2}$ is chosen so as to give an asymptotic inelasticity (= total fraction of energy carried by the emitted pions) $\eta = 0.4$. This, then, reproduces the average number

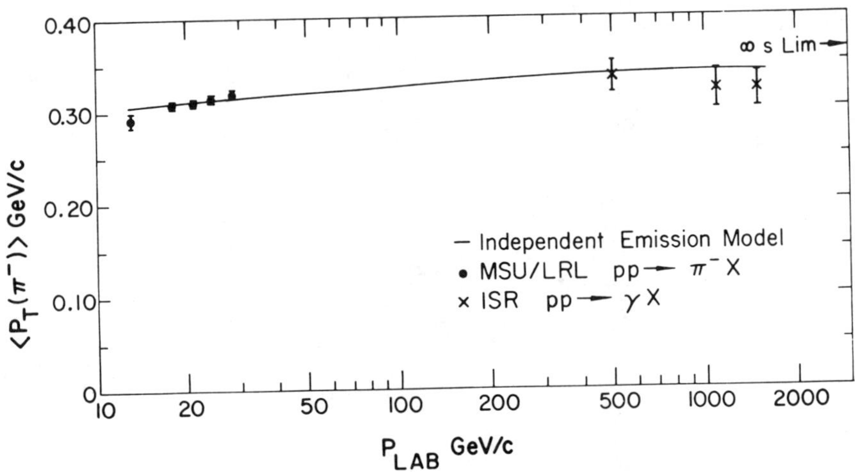

Fig. 17.2. Average transverse momenta. Taken from ref. 15.

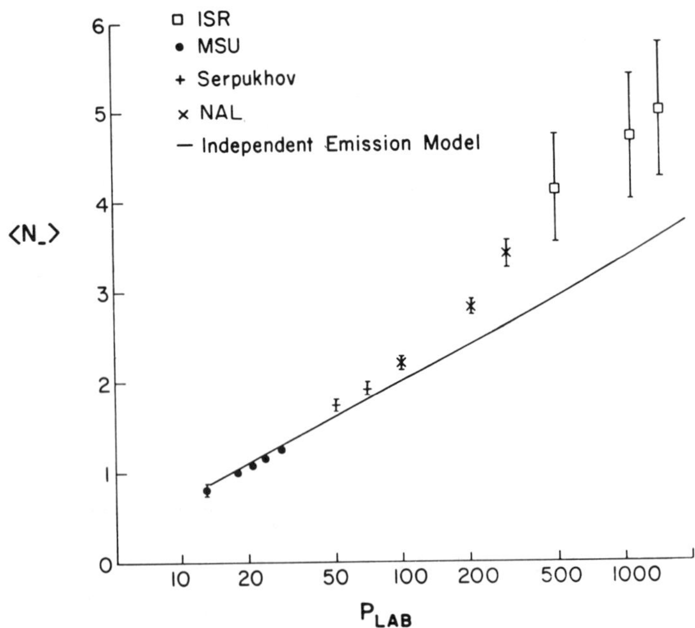

Fig. 17.3. Comparison of data with the expected variation of the average number of π^- mesons in an uncorrelated jet model. Taken from ref. 15.

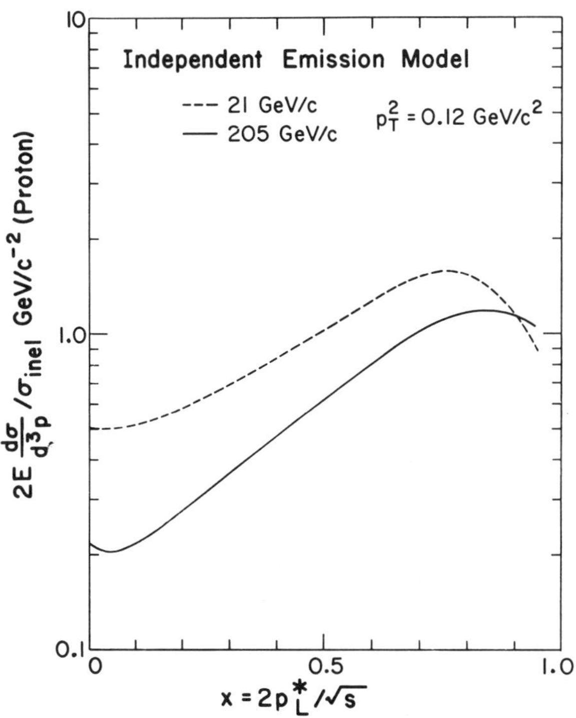

Fig. 17.4. Inclusive proton distribution in the model of ref. 15.

of π^- in the range of $s<60$ GeV^2 but fails in the NAL and ISR regions (see fig. 17.3). The proton inclusive distribution does not scale in this energy range (see fig. 17.4) and obviously misses the diffraction peak of fig. 6.12, and the pion distributions have more or less the right trends. The multiplicity distribution in the NAL range is, of course, very different from the Poisson distribution of this model, which is due to the fact that no diffractive component has been introduced here.

In the above example we see that in practical cases one tries to fit the data with explicit structure in q_T rather than the approximate longitudinal forms that we discussed above. The ensuing integrals are quite complicated and some approximations must be used to estimate them. Let us give some simple examples. A closed form for phase space formulas can be written easily in the case of Poisson distributions.[14)] Thus if

$$h_n(P) = \frac{e^{-\bar{n}}}{n!}\int \prod_i \frac{d^3q_i}{\omega_i} |g(q_i)|^2 \delta^{(4)}(P-\sum_{i=1}^{n} q_i) \tag{17.13}$$

where $P^2=s$ and $g(q_i)$ is the function specifying the distribution of the particle with momentum q_i, we can write

$$h(P) = \frac{1}{(2\pi)^4}\int d^4x\, e^{-iP\cdot x} \cdot \cdot \exp\left\{\int \frac{d^3q}{\omega} |g(q)|^2 (e^{iq\cdot x} - 1)\right\} \tag{17.14}$$

and obtain

$$h(P) = \sum_n h_n(P) = \sum_n e^{-\bar{n}} \frac{\bar{n}^n}{n!} \tilde{h}_n(P) \tag{17.15}$$

where now

$$\int d^4P\, \tilde{h}_n(P) = 1 \text{ and } \bar{n} = \int \frac{d^3q}{\omega} |g(q)|^2. \tag{17.16}$$

The closed form may be elegant (it is a natural result in the study of coherent states)[14)] but does not simplify the calculation. The function $\tilde{h}_n(P)$ can be approximated by using the central limit theorem, in which the distributions g are approximated by Gaussians in $\vec{q}^2$. In the relativistic situation discussed here this approximation takes the form

$$\tilde{h}_n(P) \approx \frac{1}{n^2} \frac{\sqrt{\det \eta}}{4\pi^2} \cdot$$

$$\cdot \exp\left\{-\frac{1}{2n} \eta_{\mu\nu}(P^\mu - n\bar{q}^\mu)(P^\nu - n\bar{q}^\nu)\right\} \qquad (17.17)$$

where we have used

$$\bar{q}_\mu = \frac{1}{\bar{n}} \int \frac{d^3q}{\omega} q^\mu |g(q)|^2$$

and

$$\frac{\eta_{\mu\nu}}{\bar{n}} \int (q^\nu - \bar{q}^\nu)\ (q^\sigma - \bar{q}^\sigma)|g(q)|^2 \frac{d^3q}{\omega} = \delta_\mu{}^\sigma. \qquad (17.18)$$

This result was obtained by Van Hove,[16)] and similar expressions were analyzed in detail by Lurcat and Mazur.[17)] The expression given above is suitable in the case of spherically symmetric distribution in the c.m., e.g., if g(q) were a function of $\frac{q \cdot P}{\sqrt{s}} = \omega$. In practical applications we have to consider a function g which incorporates the transverse momentum cut-off. Obviously it will then have a q_T dependence which is markedly different from the q_L dependence. The resultant h_n functions will still depend only on s; however, the inclusive distributions, which are defined by sums over objects of the type

$$k_n(P,q) = n\frac{e^{-\bar{n}}}{n!}\frac{|g(q)|^2}{\omega}\int\prod_{i=2}^{n}\frac{d^3q_i}{\omega_i}|g(q_i)|^2 \cdot$$

$$\cdot\delta^{(4)}(P-q-\sum_{i=2}^{n}q_i), \qquad (17.19)$$

can have the required experimental dependence on q_T and q_L. In this context it is important to note that the resulting inclusive distribution may be very different from the input $|g(q)|^2$. In particular, if we were to use in (17.19) a functional form of the type $e^{-\lambda|x|}$ for $|g(q)|^2$ we would find in the resultant $k_n(P,q)$ no x dependence, since energy momentum conservation yields $\sum_i x_i=1$ for the particles travelling to the left or to the right in the c.m. In the example of (17.12) there are two nucleons which do not have any x-dependence in $F_N(p)$. Nevertheless, the resulting inclusive nucleon distribution has a nonflat structure. It does not scale and increases exponentially over a wide range of x (see fig. 17.4). This is a reflection of energy-momentum conservation. As a matter of fact, one can start with an $F_N(p)$ which has a positive exponential dependence on $|x|$ and a roughly flat input for $G_\pi(q)$ and achieve the same net effect.[15]

In all our discussions in the present chapter we took for granted the sharp cutoff in the transverse momentum distributions. There are several ways in which one may try to justify this phenomenon. One is in terms of some Gaussian shape of matter in configuration space, in the spirit of the droplet model of the previous chapter. A different way of justifying it is given by the multiperipheral approach which blames the decrease in q_T on the effect of the exchange mechanism. A third way is given by the statistical thermodynamical model which we shall discuss in the rest of this chapter.

The first such model for production of hadrons was introduced a long time ago by Fermi.[18] He viewed the collision process as the production of a single fireball in which hadronic matter exists in thermodynamic equilibrium and then decays into observable particles. Such a model obviously fails experimentally because it does not have the longitudinal momentum characteristics of the data. This approach was modified and reshaped by Hagedorn[19] and by Hagedorn and Ranft.[20] They consider multiparticle production to result from the formation of many fireballs. The longitudinal distribution is thereby introduced externally; however, the transverse distribution reflects the properties of the decay of the fireball into particles. Thus the thermodynamical approach predicts the transverse momentum distribution. More specifically, it predicts the particle density spectrum and relates the parameters of this spectrum to those of the transverse momentum distribution.

In order to sketch the derivation of a particle spectrum let us follow the approach of Frautschi.[21] He considers hadronic matter to be compounded of two or more constituents circulating freely in a box of radius $\approx 10^{-13}$ cm. If there were a single elementary boson out of which all particles were made one could then write

$$\rho(m) = \sum_{n=2}^{\infty} \left(\frac{V}{h^3}\right)^{n-1} \frac{1}{n!} \prod_{i=1}^{n} \int d^3p_i \cdot$$

$$\cdot \delta\Big(\sum_{i=1}^{n} E_i - m\Big)\ \delta^{(3)}\Big(\sum_i \vec{p}_i\Big) \tag{17.20}$$

where $\rho(m)$ is the density of states of mass m. The basic assumption in (17.20) is that each constituent has a density of states $V\frac{d^3p}{h^3}$ and one counts the density of levels with center of mass at rest. Now one can introduce a bootstrap model of hadrons in which one starts with a density of states $\rho_{B_iS_iQ_i,in}(m)$ for a specified value of baryon number, strangeness and charge, and defines

$$\rho_{BSQ,out}(m) = \sum_{n=2}^{\infty} \left(\frac{V}{h^3}\right)^{n-1} \frac{1}{n!} \prod_{i=1}^{n} \int dm_i \cdot$$

$$\cdot \sum_i \rho_{B_iS_iQ_{i,in}}(m_i)\ \ d^3p_i\ \ \delta\Big(\sum_{i=1}^{n} E_i - m\Big)\cdot$$

$$\cdot \delta^3\Big(\Sigma \vec{p}_i\Big)\delta\Big(\sum_i B_i - B\Big)\delta\Big(\sum_i S_i - S\Big)\delta\Big(\sum_i Q_i - Q\Big). \tag{17.21}$$

The bootstrap condition is that for high values of m

$$\rho_{BSQ,out}(m) \to \rho_{BSQ,in}(m). \qquad (17.22)$$

This can be achieved if[21]

$$\rho_{BSQ}(m) \sim c_{BSQ}\, m^{a'}\, e^{bm} \qquad (17.23)$$

and the total density of states becomes

$$\rho(m) = \sum_i \rho_{B_i S_i Q_i}(m) \sim c\, m^a\, e^{bm}. \qquad (17.24)$$

Frautschi finds that $a<-\frac{5}{2}$. Hagedorn, who formulates things in a somewhat different language and effectively includes also n=1 in the sum of (17.21), obtains $a\leq-\frac{5}{2}$ and in practical applications uses $a=-\frac{5}{2}$. In fig. 17.5 we show his fits to experimental data, from which he concluded that $b^{-1}=T_o=160$ MeV.

It is interesting to note that a similar formula for the growth of level density with mass also occurs in dual models.[22] Factorization of the N-point Veneziano representation even yields a similar set of possibilities for a; namely, $a=-\frac{5}{2},-3,-\frac{7}{2}$, etc. Frautschi explains that this result is not accidental since the basic formulas of the bootstrap approach essentially provide one resonance for each scattering state, and this is

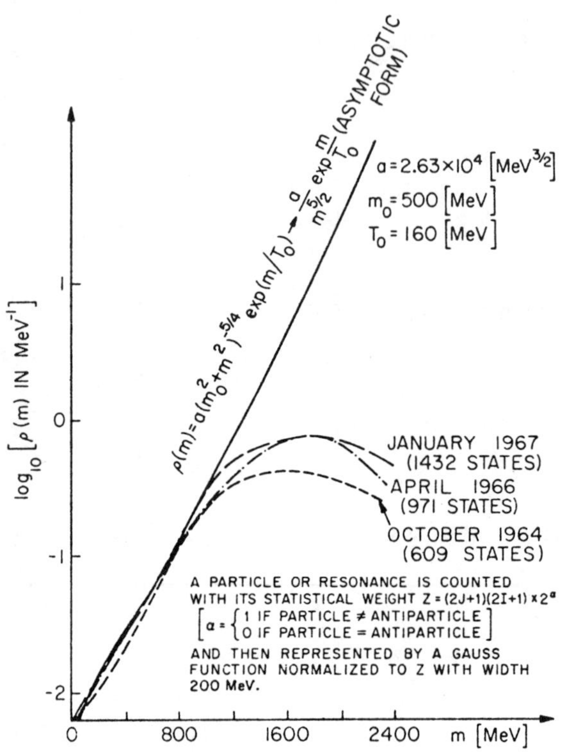

Fig. 17.5. The smoothed experimental mass spectrum as it developed from 1964 to 1967, compared with the function $\rho(m)=a(m^2+m_0{}^2)^{-5/4}\exp(m/T_0)$, which has the asymptotic form required by Hagedorn's theory. The figure is taken from ref. 19.

a natural consistency condition in dual models.

If the energy density of hadronic matter is given by eq. (17.24) then an interesting conclusion follows for hadronic matter in equilibrium. In such a state one can calculate the average energy from

$$\overline{E} = \int_0^\infty dm\ \rho(m)\ m\ e^{-m/T} \Big/ \int_0^\infty dm\ \rho(m)\ e^{-m/T}, \tag{17.25}$$

and this expression diverges for $T>T_o$. Hence the "temperature" of the hadronic matter is bounded by 160 MeV. Physically it means that as the energy increases it goes into the creation of massive particles rather than into raising the kinetic energy of the existing ones. Viewing the emitted hadron at a certain rapidity y as coming from such a hadronic equilibrium one, therefore, expects its transverse momentum distribution to be bounded by e^{-m_T/T_o}. This, in fact, turns out to be a very good guess of the actual transverse momentum dependence, which may indeed be written as $e^{-m_T/T}$ with T usually slightly lower than $T_o=160$ MeV.

The transverse momentum distribution is the single big success of the thermodynamical model. In actual fits to the data one has to introduce phenomenological distributions of the type

$$\omega\frac{d\sigma}{d^3q} = \int_{-\frac{Y}{2}}^{\frac{Y}{2}} dy\ F(Y,y)\left[\exp(\omega_y/T(y))\pm 1\right]^{-1} \quad (17.26)$$

where ω_y denotes the energy of the outgoing particle in a frame centered at rapidity y. F(Y,y) is the phenomenological function that describes the rapidity distribution (a somewhat different variable is used in the original work of ref. 19) and incorporates all other features of inclusive production.

There have also been experimental attempts to isolate and study the thermodynamic character

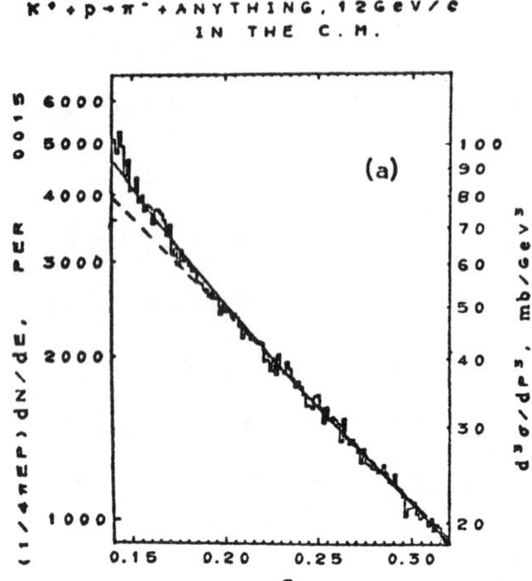

Fig. 17.6. The quantity $\frac{d^3\sigma}{d^3p}$ for a selected sample of (K^+p,π^-) is plotted vs. the c.m. energy. Taken from ref. 23.

of the distribution. In fig. 17.6 we show the results for the energy dependence of a sample of π^- produced with c.m. energy less than 320 MeV in (K^+p,π^-) at $s=24$ GeV^2. The fit to the data corresponds to a Bose-Einstein distribution $(e^{E/T}-1)^{-1}$ with T=143 MeV. It is interesting to observe that this is significantly different from a simple exponential, represented in the figure by a dashed line.

These experimental data seem to show the thermodynamical properties of a fireball which is located at the c.m. origin. In Hagedorn's model one expects to find a distribution of fireballs all along the rapidity axis. This coincides with the expectation of a plateau in rapidity of other models. Data at machine energies of $s \lesssim 60$ GeV^2 show characteristically a Gaussian rapidity distribution. The latter is actually expected in the Landau hydrodynamical model that was recently revived in ref. 24. This model uses a single fireball to account for multiparticle production (but for leading particle effects) and leads to small deviations from scaling. Recent ISR data seem to prefer a plateau to a Gaussian structure in rapidity. Forthcoming detailed data may further test this question. In particular, it will also be interesting to see if results like those of fig. 17.6 hold in the central region of ISR experiments.

18
Diffractive Models

Inclusive reactions have characteristic features which vary little over a wide range of incoming energies. This fact motivates an approach in which one attributes the main bulk of the production process to diffractive production,[25,26] as shown in fig. 18.1.

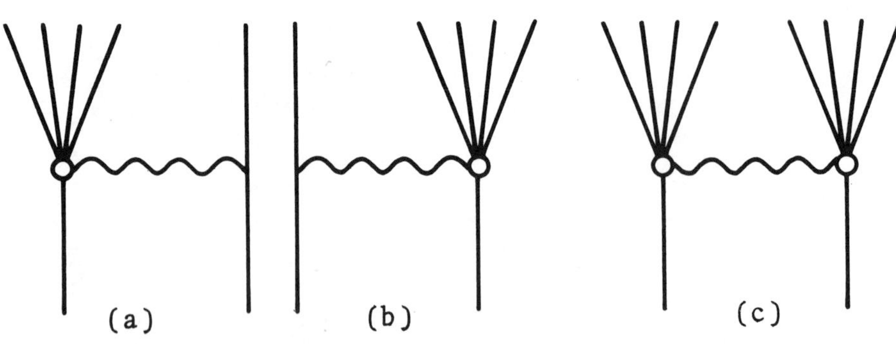

Fig. 18.1

The quantity produced is called a fireball or a nova, and the three different modes shown in fig. 18.1 correspond to single nova production, from either the projectile (a) or the target (b), and double nova production (c). The exchanged object is a Pomeron, as is appropriate for diffractive processes.

In chapter 4 we have discussed two-body diffraction dissociation, leading to the production of resonances. The characteristic cross sections were very small and the sum of all resonances listed there did not amount to more than several percent of the total cross section. As one increases the mass of the diffractively produced resonances one quickly runs into a continuum region and the first question to be settled in constructing a model of diffractive production is how to extrapolate from resonance production to continuum production.

In a Mueller approach one associates fig. 18.1(a) with a PPR vertex, since the sum over resonances should be dual to a normal Reggeon, as shown diagramatically in fig. 18.2.

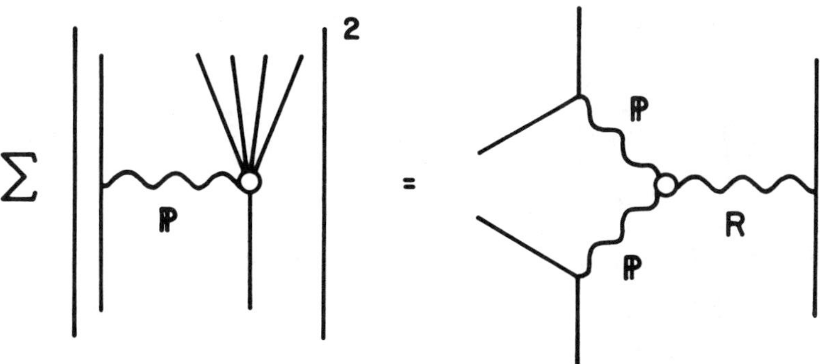

Fig. 18.2

The PPR formula for the diffractive cross section is

$$s\frac{d\sigma}{dt\ dM^2} = \frac{f(t)}{s}\left(\frac{s}{M^2}\right)^{2\alpha_P(t)}(M^2)^{\bar{\alpha}(o)}. \qquad (18.1)$$

This formula can be viewed in either one of two ways. It describes the x distribution of the leading particle near x=1; alternatively it specifies the invariant mass distribution of all the other particles produced. Hence we can view it as a quasi two-body cross section for the production of a nova of mass M:

$$\frac{d\sigma}{dt\ dM} = 2f(t)\ s^{2\alpha_P(t)-2}(M^2)^{\bar{\alpha}(o)-2\alpha_P(t)+\frac{1}{2}}.$$

(18.2)

An appropriate choice of $\bar{\alpha}$ for the PPR formula is $\bar{\alpha}(o)=\frac{1}{2}$. This leads to a cross section which decreases like M^{-2}. The contribution of a diagram like a or b in fig. 18.1 to the total diffractive cross section can be written as

$$\sigma_\alpha = \int \rho_\alpha(M)dM \qquad \rho_\alpha(M) = \int \frac{d\sigma}{dt\ dM}\,dt \qquad (18.3)$$

where α labels beam or target, depending on which has been excited. $\rho(M)$ is called the excitation spectrum. For high values of M, $\rho_\alpha(M)$ should be determined by (18.2). If one associates the diffractively produced resonances, discussed in chapter 4, with an already falling excitation spectrum, then the diffractive cross section cannot account for a large part of the inelastic events. Instead, Jacob and Slansky[26)] suggested the following form:

$$\rho_\alpha(M) = c_\alpha \frac{\exp\left[-\beta_\alpha/(M-M_\alpha)\right]}{(M-M_\alpha)^2} \qquad (18.4)$$

where M_α is the mass of the particle that undergoes diffractive excitation, and c_α and β_α are parameters characteristic of the nova. This form (18.4) has the required asymptotic M^{-2} falloff and peaks at $M_{max}=M_\alpha+\frac{1}{2}\beta$. A choice of $M_{max}=2$ GeV for proton novas and 1.3 GeV for pion novas leads to a successful representation of the data[26)] in the low s range, $s\lesssim 60$ GeV^2.

Let us turn now to the prediction of inclusive distributions in this model. A nova is assumed to decay into lower mass hadrons and eventually to cascade into the observed mesons and baryons. If one thinks of the nova as a kind of heavy resonance one would be inclined to assume that the number of particles that result from its decay will be proportional to the nova mass, $n\propto M$. This would mean that the available energy goes into production of more pions rather than giving them higher energies, consistent with the discussion at the end of the previous chapter. Since $\frac{d\sigma}{dM}\propto M^{-2}$ we are inevitably led to

$$\sigma_n \propto n^{-2}. \qquad (18.5)$$

The production cross section σ_n has, of course, to be cut off at the mass $M=\frac{1}{2}\sqrt{s}$, which therefore leads to the predictions

$$\langle n \rangle \propto \ln s, \text{ and } C_n \propto (\sqrt{s})^{n-1} \text{ for } n>1. \quad (18.6)$$

The logarithmic increase of $\langle n \rangle$ is, of course, a welcome feature. However, $C_2 \sim \sqrt{s}$ implies $f_2 \sim \sqrt{s}$, and this may very well be too rapid an increase of the two particle correlations. Two-particle distributions are evidently a sensitive test of the model, since all parameters are determined by the fit to the single-particle inclusive distributions, so that no further freedom remains in the two-particle situation.

Let us elaborate further on the decay properties of the nova. In its rest frame, the nova of mass M is assumed to decay into particles of type i isotropically, according to the form

$$n_i(M) \exp -\left(\frac{\vec{q}\,'^2}{K^2}\right) \quad (18.7)$$

where $n_i(M) \propto M$, $K \approx 300$ MeV, and $\vec{q}\,'$ denotes the momentum of the secondary particle in the nova rest frame. The choice of K is determined by the transverse momentum cutoff. The connection between q'_L and x can be calculated:

$$x = \frac{2}{M\sqrt{s}}\left[q_N(m_T{}^2+q_L'^2)^{\frac{1}{2}} + q_L'(M^2+q_N{}^2)^{\frac{1}{2}}\right] \quad (18.8)$$

where $m_T^2 = m_i^2 + \vec{q}_T'^2$, M is the nova mass and q_N is its longitudinal momentum in the center of mass frame. For x away from 0 one can approximate (18.8) and write

$$\frac{d\sigma}{dx\ dq_T^2} \approx \frac{1}{2}\int dM\rho(M)Mn_i(M)\ \exp\left\{-\frac{x^2M^2+4q_T^2}{4K^2}\right\} \tag{18.9}$$

which gives the contribution of either the beam or target nova to the inclusive distribution of particle i. In this approximation one finds a scaling behavior for the distribution[26)] in the form of a Gaussian in x (roughly e^{-10x^2} and e^{-5x^2} for baryon and meson fragmentation respectively) and the expected cutoff in q_T. The approximation fails for $x \lesssim m_i/M$. The point $x = \frac{m_i}{M}$ is where the distribution of the secondaries will peak. In order to obtain pionization products one needs the M^{-2} tail of (18.3).

Another interesting aspect of the model is the spectrum of the leading particles. To be specific, we may look at the distribution (pp,p). The diagram (18.1a) will lead to a peak in the neighborhood of $x \approx 1$ which will get sharper as the energy increases (nonscaling behavior because $\bar{\alpha} = \frac{1}{2}$). Figs. 18.1b and 18.1c lead to continuous distributions which decrease in the neighborhood of $x \approx 1$. Their combined effects are shown in fig. 18.3 where the three contributions are denoted by I, II, and III respectively and their sum

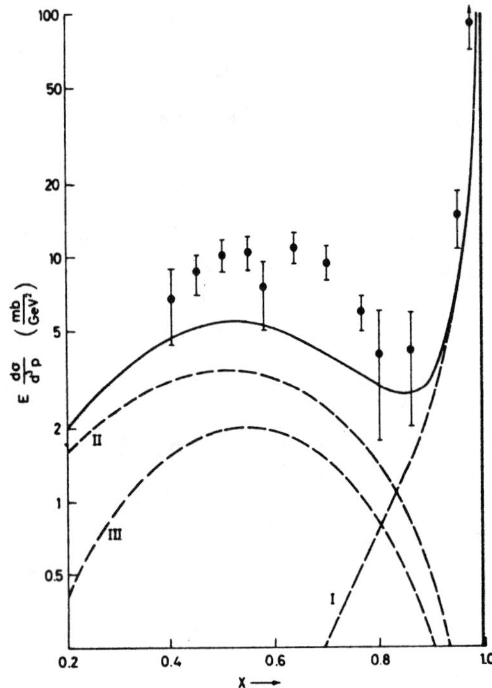

Fig. 18.3. ISR data are compared with nova model calculations of (pp,p). Taken from ref. 27.

is represented by the solid line. This figure is taken from ref. 27 in which the diffractive model is compared with the earliest data available from the ISR. The (pp,p) data are taken at θ=55 mrad and s=440 (GeV)^2. The parameters of the model ascribe 29% of σ_I to nondiffractive effects and have about equal weights for the three possibilities of fig. 18.1. The deep valley in the (pp,p) distribution is observed in the ISR and is absent in the lower energy data. The authors of ref. 27 conclude, therefore, that such a model may have some chance of success at very high energies. However, recent data show that the structure seen at the ISR has a scaling, or triple Pomeron, component (see the discussion in chapter 6); hence its energy behavior cannot be successfully described by a diffractive picture.

The model suffers from several other defects. The most noticeable one is, of course, the $\sigma_n \propto n^{-2}$ prediction. The failure of this prediction is clearly brought out in fig. 18.4, in which we see $n^2\sigma_n$ plotted vs. n^4. The reason for plotting the data in this fashion is to show that experimentally the high energy topological cross sections behave approximately as[28)]

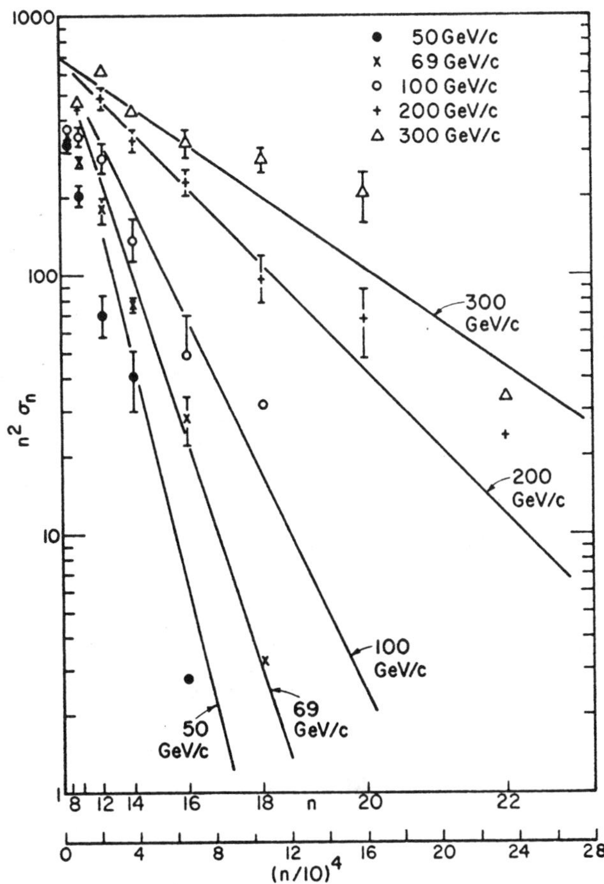

Fig. 18.4. Prong distribution in pp collisions are compared with the fit of eq. (18.10). Taken from ref. 28.

$$n^2\sigma_n = 700 \exp\left[-\frac{70}{s}\left(\frac{n}{10}\right)^4\right] \text{mb}. \qquad (18.10)$$

The lines in fig. 18.4 correspond to this fit.

Such a behavior can be understood in the diffractive model if one adds a damping factor e^{bt} in the production of the two fireballs. This is a plausible factor since such a t dependence is actually seen in two-body diffraction dissociation (see chapter 4) and is therefore a natural choice for f(t) in (18.1). In a process such as (pp,p), where a missing mass M is produced, one finds at large s that the minimum value of t is

$$-t_{min} \approx M^4 \, m_p^{\,2}/s^2. \qquad (18.11)$$

Double nova production of novas having masses M_1 and M_2 has the minimum -t value

$$-t_{min} \approx M_1^{\,2} M_2^{\,2}/s. \qquad (18.12)$$

Hence these reactions should be damped by a factor of $e^{bt_{min}}$. Choosing $M_1 \approx M_2$ one finds

$$\ln n^2\sigma_n \propto -n^4/s^2, \qquad \text{single nova}$$

and $$\ln n^2\sigma_n \propto -n^4/s, \qquad \text{double nova.} \qquad (18.13)$$

In view of the fit of (18.10) in fig. 18.4

it may therefore seem that if diffractive production of two novas is dominant over production of one, then the model can in fact account for the behavior of the topological cross sections. This, however, contradicts earlier fits[26)] to inclusive distributions at $s \lesssim 60$ GeV^2; these required a 2:1 ratio for single:double nova production. We note also that a pure double nova picture cannot produce the peak in the leading particle distribution of fig. 18.3; and the worst aspect of this approach is that the strong damping of the production of high mass novas strongly reduces the production of low x pions and spoils the fits to inclusive distributions in the central region. The accumulating evidence thus shows that a purely diffractive model cannot account for the high energy data.

Let us return now to our starting point in fig. 18.2. Although we began with a generalization of the triple Regge approach, we made the very crucial assumption that the number of secondary particles was proportional to the mass of the nova in eq. (18.4), and we would like to emphasize that this contradicts the nature of the usual models that lead to a triple Regge formula. For example, in a multiperipheral model one can also expect a diffractive component of the type shown in fig. 18.2, but the number of secondaries included in M will be only

$$n \approx c \ln M^2. \qquad (18.14)$$

Incorporating this into a Pomeron-Pomeron-Regge vertex we find from (18.2) that since $dn=2cdM/M$,

$$\sigma_n \sim \frac{d\sigma}{dn} \propto M\frac{d\sigma}{dM} \propto e^{-n/c} \text{ for } \bar{\alpha} = \frac{1}{2}, \tag{18.15}$$

namely a strong cutoff in the number of particles produced diffractively. If we consider instead a triple Pomeron vertex we find a distribution which is essentially constant, though the details depend on what we assume for the triple Pomeron vertex function. If we assume it to vanish linearly at $t=0$,

$$\frac{d\sigma}{dt\ dM^2} = \left(\frac{s}{M^2}\right)^{2\alpha(t)-2} f(t), \quad f(t) \sim -te^{at}, \tag{18.16}$$

then we obtain

$$\sigma_n \approx M^2 \frac{d\sigma}{dM^2} \propto \frac{1}{\left[a + 2\alpha'(\ell n\ s - \frac{n}{c})\right]^2}, \tag{18.17}$$

while $f(t) \sim e^{at}$ would result in a single power in the denominator of (18.17). In eq. (18.17) we see that for every finite n, σ_n decreases like $(\ell n\ s)^{-2}$; however, it peaks towards the upper limit and therefore has still the property that

$$\langle n \rangle = c\ \ell n\ s. \tag{18.18}$$

Both these PPR and PPP distributions are, of

course very different from the nova model; nevertheless, they deserve the name diffractive production. The observation of diffractive production in the (pp,p) inclusive distribution obviously means that a diffractive mode is present. This does not, however, mean that diffractive production is the principal mechanism of particle production, though it may be responsible for a fair part of it. In chapter 3, for example, we saw that a simple model which includes both multiperipheral and diffractive modes requires a ratio of about 4:1 between the two in order to account for the observed behavior of inelastic cross sections. We shall return to such hybrid models in chapter 20.

19

Probability Distributions of Pions

In many of the models that we have discussed, such as the multiperipheral models in part IV and the diffractive models in chapter 18, we presented expressions for partial cross sections σ_n without specifying the types of particles produced. We would now like to go one step further and consider models of both the multiperipheral (i.e., Poisson) or nova ($\sigma_n \propto n^{-2}$) type distributions for different charge states of pions. This is of practical interest since there already exist data which are sensitive to the details of the models.

The data that we refer to concern the variation of $\langle n_o \rangle_-$, namely, the average number of π^o produced for a fixed number of π^-. This quantity is relatively easily measured in high-energy experiments, since it suffices to limit oneself to a specific number of prongs and ask for the average yield of γ-rays. Neglecting production of particles other than pions and nucleons, one finds in pp reactions that the number of π^- is $n_-=\frac{1}{2}(n_{ch}-2)$ and, of course, $n_\gamma=2n_o$. By looking at the variation of $\langle n_o\rangle_{n_-}$ with n_- one tests correlations in the system of the particles produced. Experimentally one observes a linear increase of $\langle n_o\rangle_{n_-}$ with n_- as shown in fig. 19.1. Our first purpose will be to understand this rise in different production models.

All the models that we are going to discuss[30,31] have two common characteristics: First, they involve only production of pions either singly or in clusters, but other effects, such as

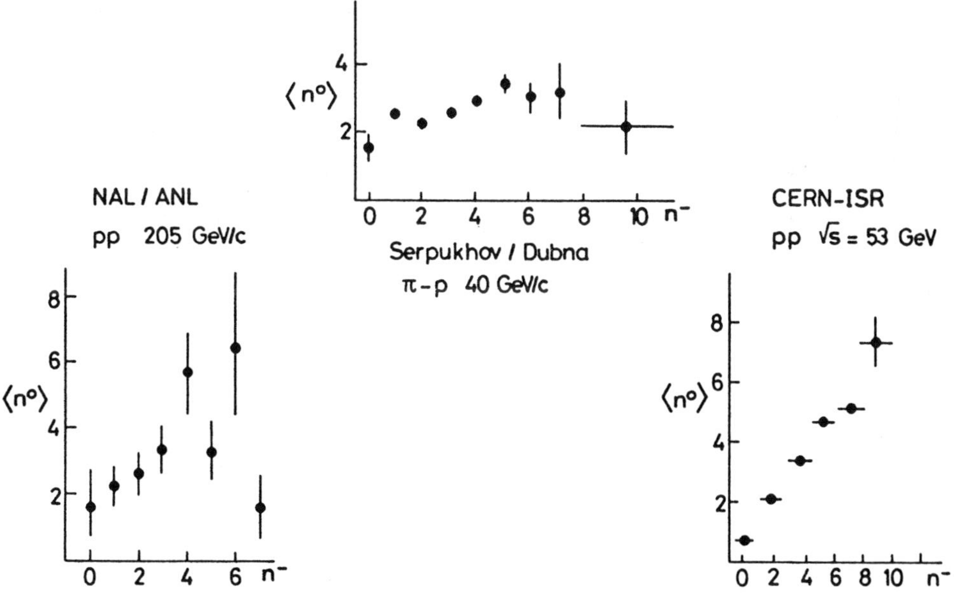

Fig. 19.1. The mean number of π^o as a function of the observed number of charged particles. Data from a Dubna collaboration at 40 GeV/c, the Argonne-NAL experiment at 200 GeV and an ISR experiment are combined together. The observed ISR multiplicities are limited to a particular solid angle. Taken from ref. 29.

K and $\bar{p}$ production are neglected. In addition, the models lead asymptotically to the same average number of neutral, negative, or positive pions,

$$\langle n_o \rangle = \langle n_+ \rangle = \langle n_- \rangle, \tag{19.1}$$

although most models deviate from eq. (19.1) at finite energies.

We will classify the models according to two features - the multiplicity distribution of clusters and the isospin properties of the clusters. Thus we will discuss σ, π and ρ models, which just means production of I=0 pion pairs, single pions, and I=1 pairs respectively. In addition we will assume that they fall into the category of either a Poisson or a power distribution, corresponding to simplified characteristics of the multiperipheral and nova models.

Let us start with the σ-model. If we designate by n the number of $\pi^+\pi^-$ pairs and by k the number of $\pi^o\pi^o$ pairs we find

$$n_- = n \qquad n_o = 2k \tag{19.2}$$

and a simple Poisson distribution will be

$$P_{\sigma 1}(n_-,n_o) = e^{-z} \frac{\left(\frac{2}{3}z\right)^n}{n!} \frac{\left(\frac{1}{3}z\right)^k}{k!} \tag{19.3}$$

where we have assigned a $\frac{2}{3}$ probability to the $\pi^+\pi^-$

configuration and a $\frac{1}{3}$ probability to the $\pi^o\pi^o$ one, as is appropriate for an I=0 configuration. z is the average number of σ-clusters produced. We can now rewrite eq. (19.3) as a Poisson distribution $P_{\sigma 2}$ which characterizes the division into n and k once N is given:

$$P_{\sigma 1} = e^{-z}\,\frac{z^N}{N!}\,P_{\sigma 2}, \quad P_{\sigma 2} = \binom{n+k}{n}\left(\frac{2}{3}\right)^n\left(\frac{1}{3}\right)^k,$$

where

$$N = n + k = n_- + \frac{n_o}{2}\,. \tag{19.4}$$

If we now want to change from a Poisson law to a power law, e.g., take N^{-2} as appropriate for a nova model, we find

$$P_{\sigma 3} = \frac{c_\sigma}{N(N-1)}\,P_{\sigma 2}. \tag{19.5}$$

where we have written N(N-1) instead of N^2 just for mathematical convenience. c_σ is a normalization constant.

Let us now define the π and ρ models. Here we discuss the production of charged particles (or clusters of particles); therefore, a simple Poisson distribution is not appropriate since it does not conserve charge. Let us therefore impose charge conservation on a Poisson distribution, defining

$$P_{\pi 1}(n_-,n_o) = \frac{e^{-z/3}}{I_o\left(\frac{2}{3}z\right)}\,\frac{\left(\frac{z}{3}\right)^{2n}\left(\frac{z}{3}\right)^{n_o}}{n!n!n_o!}, \quad n=n_-, \tag{19.6}$$

and

$$P_{\rho 1}(n_-,n_o) = \frac{e^{-z/3}}{I_o\left(\frac{2}{3}z\right)} \frac{\left(\frac{z}{3}\right)^{2k}\left(\frac{z}{3}\right)^{n-k}}{k!k!(n-k)!},$$

$$n=n_-, \quad n_o=2k. \tag{19.7}$$

In eq. (19.6) $n=n_-=n_+$ is the number of positive or negative pions. In eq. (19.7) k is the number of positive or negative ρ-clusters and (n-k) is the number of neutral ones. Since $\rho^{\pm}\to\pi^{\pm}\pi^{o}$ and $\rho^{o}\to\pi^{+}\pi^{-}$ one finds that $n_o=2k$ and $n_-=n-k+k$.

Eqs. (19.6) and (19.7) are appropriate for describing the production of an uncorrelated cloud of particles of total charge zero. Such expressions were displayed before in eqs. (17.9) and (17.10), where we pointed out that taking account of charge conservation in this manner also approximately satisfies overall isospin conservation. Changing from an overall charge of zero to a charge of one or two does not make much difference in the final result. Also modifications due to energy-momentum conservation are relatively unimportant and will not be considered here. We refer the reader to ref. 30 for further details.

For high values of n_- and n_o we can rewrite eq. (19.6) and eq. (19.7) in the asymptotic forms

$$P_{\pi 1}(n_-,n_o) \to C\frac{z^N}{N!}\binom{N}{n_o}\left(\frac{2}{3}\right)^{2n_-}\left(\frac{1}{3}\right)^{n_o}, \quad N=2n_-+n_o \tag{19.8}$$

$$P_{\rho 1}(n_-,n_o) \to C\frac{z^N}{N!}\frac{(n+k)!}{k!k!(n-k)!}\left(\frac{1}{3}\right)^{k+n},$$

$$N=n+k=n_-+\frac{n_o}{2}, \tag{19.9}$$

which exhibit the Poisson characteristics. N is the total number of pions in eq. (19.8) and the total number of ρ-clusters in eq. (19.9). We can now read off the analogues of $P_{\sigma 2}$, namely

$$P_{\pi 2} = \frac{(2n_- + n_o)!}{(2n_-)!\, n_o!} \left(\frac{2}{3}\right)^{2n_-} \left(\frac{1}{3}\right)^{n_o}, \tag{19.10}$$

$$P_{\rho 2} = \frac{(n+k)!}{k!\,k!\,(n-k)!} \left(\frac{1}{3}\right)^{k+n}, \quad n=n_-, n_o = 2k, \tag{19.11}$$

and, switching to an N^{-2} distribution, one can define the nova type models

$$P_{\alpha 3} = \frac{c_\alpha}{N(N-1)} P_{\alpha 2}, \qquad \alpha = \sigma, \pi, \rho. \tag{19.12}$$

Let us turn now to the calculation of $\langle n_o \rangle_-$. This can be simply done by defining a generating function

$$F(n_-, x) = \sum_{n_o} P(n_-, n_o)\, x^{n_o} \tag{19.13}$$

from which one obtains

$$\langle n_o \rangle_- = \frac{\partial}{\partial x} \ln F(n_-, x) \Big|_{x=1} . \tag{19.14}$$

Straightforward computation leads to the following generating functions (with arbitrary normalization and using $n = n_-$)

$$F_{\sigma 1} = \frac{1}{n!}\left(\frac{z}{3}\right)^n e^{\frac{z}{3}x^2}, \quad F_{\sigma 2} = \left(\frac{1}{3}\right)^n \left(1-\frac{x^2}{3}\right)^{-1-n},$$

$$F_{\sigma 3} = \frac{1}{n(n-1)}\left(\frac{1}{3}\right)^n \left(1-\frac{x^2}{3}\right)^{1-n}, \tag{19.15}$$

$$F_{\pi 1} = \frac{1}{n!n!}\left(\frac{z}{3}\right)^{2n} e^{\frac{z}{3}x}, \quad F_{\pi 2} = \left(\frac{2}{3}\right)^{2n}\left(1-\frac{x}{3}\right)^{-1-2n},$$

$$F_{\pi 3} = \frac{1}{2n(2n-1)}\left(\frac{2}{3}\right)^{2n}\left(1-\frac{x}{3}\right)^{1-2n}, \tag{19.16}$$

$$F_{\rho 1} = \frac{1}{n!}\left(\frac{z}{3}\right)^n L_n\left(-\frac{z}{3}x^2\right), \quad F_{\rho 2} = \left(\frac{1}{3}\right)^n P_n\left(1+\frac{2}{3}x^2\right),$$

$$F_{\rho 3} = \frac{1}{n(n-1)}\left(\frac{1}{3}\right)^n P_n^{(o,-2)}\left(1+\frac{2}{3}x^2\right) \tag{19.17}$$

where L_n, P_n and $P_n^{(o,-2)}$ denote the Laguerre, Legendre and Jacobi polynomials, respectively. It is now easy to see that

$$\langle n_o\rangle_- = \frac{2}{3}z \text{ for } P_{\sigma 1}, \quad \langle n_o\rangle_- = n_- - 1 \text{ for } P_{\sigma 3} \tag{19.18}$$

and $$\langle n_o\rangle_- = \frac{z}{3} \text{ for } P_{\pi 1}, \langle n_o\rangle_- = n_- - \frac{1}{2} \text{ for } P_{\pi 3}. \tag{19.19}$$

In both of these cases, a multiperipheral model leads to independent production of particles and, therefore, $\langle n_o\rangle_-$ is constant, contrary to experimental observation. If, however, one turns to nova-type models, a linear increase of $\langle n_o\rangle_-$ is obtained. Does this mean that a multiperipheral approach is ruled out? Not really, since a Poisson production of ρ-clusters leads to an increasing $<n_o>_-$.[32] Using eq. (19.17) we find

$$<n_o>_- = 2n_- \left(1 - \frac{L_{n-1}\left(-\frac{z}{3}\right)}{L_n\left(-\frac{z}{3}\right)}\right) \text{ for } P_{\rho 1}. \qquad (19.20)$$

This function increases for low values of n_- and levels off later:

$$<n_o>_- \approx 2n_- \qquad \text{for } n_- << \tfrac{1}{3}z$$

and $$<n_o>_- \approx 2\left(\frac{zn}{3}\right)^{\frac{1}{2}} \qquad \text{for } n_- >> \tfrac{1}{3}z \qquad (19.21)$$

Changing from $P_{\rho 1}$ to $P_{\rho 3}$ (from multiperipheral to nova ρ-production) one obtains

$$<n_o>_- \approx n_- - 1 \qquad \text{for } P_{\rho 3} \qquad (19.22)$$

Here we quote an approximate result, instead of an exact form in terms of $P_n^{(o,-2)}\left(\frac{5}{3}\right)$, in order to show the similarity to $P_{\sigma 3}$ and $P_{\pi 3}$. Summarizing the situation we may say that the $P_{\rho 1}$ model and all $P_{\alpha 3}$ models, for any α can account for the experimental data of fig. 19.1.

It is interesting to understand the reason for these results. We can do so[31] through a simplified model for $P(n_-, n_o)$. We want this model to account for the qualitative features of the correlations between neutral and charged pions.

These can be understood as the result of two competing mechanisms: (a) The total number of pions is concentrated around an average value which is determined by the total energy only. Hence an increase of n_- should eventually lead to a decrease of n_o. (b) The number of neutral pions tends to be equal to the number of negative pions since the overall isospin of the system is limited. Hence an increase of n_- should lead to an increase of n_o.

The interplay between these two mechanisms can be exemplified by using Gaussian distributions for the two effects: Let us take

$$P(n_-,n_o) = \frac{3}{2\pi\sigma_1\sigma_2} \exp\left[-\frac{(2n_-+n_o-\nu)^2}{2\sigma_1^2}\right] \cdot \exp\left[-\frac{(n_--n_o)^2}{2\sigma_2^2}\right], \tag{19.23}$$

where ν is the average total number of pions determined by the overall energy in the process and the two widths σ_1 and σ_2 reflect the strengths of the two constraints. For a fixed value of n_-, eq. (19.23) becomes a Gaussian in n_o with the following parameters:

$$\langle n_o \rangle_- = \frac{\sigma_1^2 n_- - \sigma_2^2(2n_--\nu)}{\sigma_1^2+\sigma_2^2} \tag{19.24}$$

$$\sigma^2_{n_-} = \frac{\sigma_1^2\sigma_2^2}{\sigma_1^2+\sigma_2^2} . \tag{19.25}$$

These equations lead to two important conclusions:

1. An increase of $\langle n_o \rangle_-$ with n_- follows if $\sigma_1^2 >> 2\sigma_2^2$.
2. In the region where $\langle n_o \rangle_-$ increases one should then find $\sigma_{n_-}^2 \approx \sigma_2^2$, namely the isospin constraint determines the shape of the distribution.

This analysis leads to a qualitative understanding of the results obtained above. We see that in the case of ρ pair production σ_2^2 is very small and, therefore, an increasing $\langle n_o \rangle_-$ is naturally obtained for any overall production mechanism. This is not the case for single π or σ pair production. Only when one uses a power law for the description of the overall distribution does one increase σ_1 so much that the condition $\sigma_1^2 >> 2\sigma_2^2$ is obtained. Furthermore, we see that in this region, where $\langle n_o \rangle_-$ increases with n_-, the shape of the distribution is determined by the isospin structure of the various models and not by the overall character of the distribution. It is, therefore, to be expected that the measurement of additional parameters of associated distributions can distinguish between the various models.

At this stage we may make a comparison between the different models and the Gaussian distribution function of eq. (19.23). First, let us plot the distributions $P_{\sigma 2}$, $P_{\pi 2}$, and $P_{\rho 2}$ as a function of n_o for a fixed value of n_-. This is shown in fig. 19.2 where we also add a Poisson

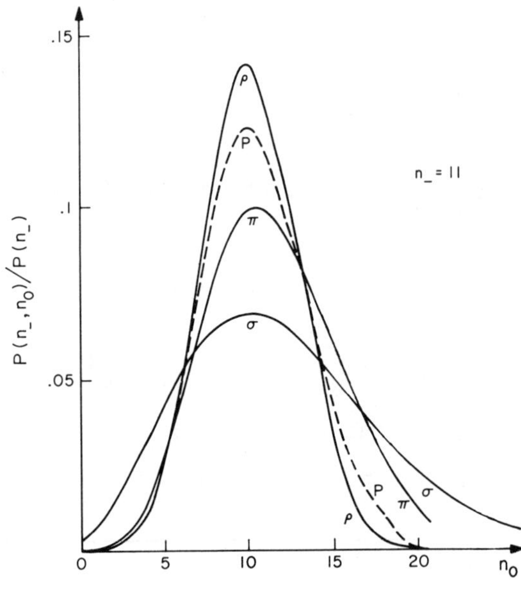

Fig. 19.2. Probability distribution of n_0 for fixed $n_-=11$ are plotted for $P_{\sigma 2}$, $P_{\pi 2}$, and $P_{\rho 2}$. The various curves are labeled σ, π, and ρ, respectively. The curve labeled P is a Poisson distribution with the same mean $\langle n_0 \rangle_-$. Taken from ref. 30.

distribution for comparison. A direct calculation leads to

$$\langle n_0 \rangle_- = n_- + 1, \quad \sigma^2_{n_-} = 3n_- + 3 \qquad \text{for } P_{\sigma 2} \quad (19.26)$$

$$\langle n_0 \rangle_- = n_- + \frac{1}{2}, \quad \sigma^2_{n_-} = \frac{3}{2}n_- + \frac{3}{4} \qquad \text{for } P_{\pi 2} \quad (19.27)$$

$$\langle n_0 \rangle_- = n_-, \quad \sigma^2_{n_-} = \frac{3}{4}n_- \qquad \text{for } P_{\rho 2} \quad (19.28)$$

We may remark that Gaussian distributions with the parameters given by (19.26)- (19.28) provide excellent numerical approximations to the distributions $P_{\alpha 2}$, thus justifying the approximation of the Guassian model.

It is clear that only $P_{\rho 2}$ leads to a form which is narrower than a Poisson distribution and therefore only it will have a negative value of $f^o_{2,n_-} = \sigma^2_{n_-} - \langle n_o \rangle_-$. All other distributions have positive f^o_{2,n_-} values. This is an important distinction to which we will return later. The values of $\sigma^2_{n_-}$ quoted in eqs. (19.26) - (19.28) are the analogues of $\sigma_2{}^2$ in eq. (19.23). In order to find $\sigma_1{}^2$ we have just to look at the Poisson distribution which is factored out in eqs. (9.4), (9.8), and (9.9). Its average as well as its variance are given by z. Note however that N designates the number of pions in eq. (9.8) and the number of pion pairs in eqs. (9.4) and (9.9). We find therefore for the three cases

$$\sigma_1{}^2 = 4z \quad \nu = 2z \quad \sigma_2{}^2 = 3n_- + 3 \text{ for } P_{\sigma 1} \tag{19.29}$$

$$\sigma_1{}^2 = z \quad \nu = z \quad \sigma_2{}^2 = \tfrac{3}{2}n_- + \tfrac{3}{4} \text{ for } P_{\pi 1} \tag{19.30}$$

$$\sigma_1{}^2 = 4z \quad \nu = 2z \quad \sigma_2{}^2 = \tfrac{3}{4}n_- \quad \text{ for } P_{\rho 1} \tag{19.31}$$

In practical applications we are usually considering the region where $n_- \approx \langle n_- \rangle$. For $P_{\sigma 1}$ one finds $\langle n_- \rangle = \frac{2}{3}z$ and at this point $\sigma_1{}^2 \approx 2\sigma_2{}^2$, which is actually the condition for no n_- dependence in eq. (19.24). For $P_{\pi 1}$ one finds $\langle n_- \rangle = \frac{z}{3}$ and a similar result follows. In both of these cases one can actually see from the definitions that no correlation between n_o and n_- should be expected. The situation is different for P where one can clearly see that $\sigma_1^2 > 2\sigma_2^2$ in the relevant region and therefore a linear rise of $\langle n_o \rangle_-$ is expected.

Turning now to the distributions $P_{\alpha 3}$ we note that they have no energy (or z) dependence and their N behavior is a very broad and smooth distribution. Therefore, one finds in all cases $\sigma_1^2 > 2\sigma_2^2$ and an increasing $\langle n_o \rangle_-$ follows, as we already have seen above.

Now we can go one step further and ask how to distinguish between the various models. A look at fig. 19.2 provides the answer - measure their width. We already saw in eq. (19.25) that in the region where $\langle n_o \rangle_-$ increases, the width of the distribution is determined by $\sigma_{n_-}^2$ and, therefore, a measurement of f^o_{2,n_-} should show the trends indicated in fig. 19.2. This is actually the case. f^o_{2,n_-} can be readily calculated from the generating functions, eqs. (19.15) - (19.17), though the equality

$$f^o_{2,n_-} = \frac{\partial^2}{\partial x^2} \ln F(n_-, x) \Big|_{x=1} \qquad (19.32)$$

and the following results are obtained

$$f^o_{2,n_-} = 2n_- - 2 \qquad \text{for } P_{\sigma 3} \qquad (19.33)$$

$$f^o_{2,n_-} = \frac{1}{4} \qquad \text{for } P_{\pi 3} \qquad (19.34)$$

and

$$f^o_{2,n_-} = n_- \left(-2+6 \frac{L_{n-1}(-\frac{z}{3})}{L_n(-\frac{z}{3})} -4 \frac{L_{n-2}(-\frac{z}{3})}{L_n(-\frac{z}{3})} \right) + 4n_-^2 \left(\frac{L_{n-2}(-\frac{z}{3})}{L_n(-\frac{z}{3})} - \frac{L_{n-1}^2(-\frac{z}{3})}{L_n^2(-\frac{z}{3})} \right) \text{ for } P_{\rho 1}. \quad (19.35)$$

Thus $P_{\sigma 3}$ predicts a strong positive correlation increasing with n_-. $P_{\pi 3}$ implies a small constant, and $P_{\rho 1}$ leads to a negative f^o_{2,n_-} which is slowly decreasing. A stronger decrease is implied by $P_{\rho 3}$:

$$f^o_{2,n_-} \approx -\frac{n}{4} + 1 \qquad \text{for } P_{\rho 3}. \quad (19.36)$$

For comparison we can mention still another possible model - the ω model - in which $\pi^+\pi^-\pi^o$ triplets are produced together. In this case one obviously has $n_o = n_-$ and, therefore,

$$\langle n_o \rangle_- = n_- \qquad f^o_{2,n_-} = -n_- \qquad \text{for } \omega \text{ model} \quad (19.37)$$

showing a strong negative correlation.

At this point we may wonder about the results for f^o_2, namely the inclusive correlation between two neutral pions. The relation between f^o_2 and f^o_{2,n_-} can be readily derived and written as a sum rule

$$\sum_{n^-} P(n_-)\left[f_2^o - f_{2,n_-}^o + \langle n_o\rangle_- \left(\langle n_o\rangle - \langle n_o\rangle_-\right)\right] = 0. \tag{19.38}$$

Since this relation involves additional terms one cannot learn the f_2^o value directly from the f_{2,n_-}^o behavior. This is clearly brought out in the ω model, in which eq. (19.38) implies just that $f_2^o = f_2^-$. The value of f_2^o depends therefore very much on the overall behavior of the distribution. Thus in the $P_{\rho 1}$ model one finds that whereas $f_{2,n_-}^o \to \frac{1}{2} - \frac{z}{3}$ in the $n_- \gg z$ region, the inclusive correlation tends to a positive value $f_2^o \to \frac{1}{2}$.

To summarize this chapter let us recapitulate the main points. We have seen that presently available data are inconsistent with a multiperipheral model of independent pion production or I=0 pion-pair production. Within the multiperipheral model the fact that $\langle n_o\rangle_-$ grows with n_- means that there is a positive correlation between the π^o and π^- particles such as can be found in ρ or ω cluster formations. Alternatively, it is possible that the main effect comes from very broad overall distributions such as the nova model. In view of the results in the previous chapter we consider the latter explanation unlikely. We expect, therefore, that if one looks at further distribution parameters, such as f_{2,n_-}^o, one will find the characteristics of multiperipheral cluster productions that we have discussed.

The various models that we describe here

are, of course, oversimplified examples. Nevertheless, one may expect that any more detailed model which incorporates all necessary constraints such as energy-momentum and isospin conservation, will still have features similar to those of one of the models that we discussed. In particular, let us note that a Poisson distribution cannot account for the observed multiplicities, and therefore long-range modifications of the multiperipheral model should be introduced. However, one may still expect that the main bulk of particle production, i.e., the distributions in the central region, have multiperipheral characteristics. It is, therefore, important to realize that the test of a parameter like f^{o}_{2,n_-} is independent of the exact nature of the overall multiplicity distribution and can therefore give us direct information about cluster formations. In addition, we may expect that future experiments will be able to select explicit regions in rapidity, e.g., $-2 \le y \le 2$, and test the correlations within these regions (summing over everything outside the region). This will provide us with further information and may help to build a consistent detailed production picture.

20

Hybrid Models

All throughout the book we have discussed various types of models but, as the reader has certainly noticed, we have always referred back to and compared with the multiperipheral model. The reason for this is the simplicity and transparency of the mathematical structure of this model. Its predictions for inclusive reactions are essentially the same as the factorizing Regge picture discussed in part II, and its theoretical basis has been investigated in part IV. It has, however, both theoretical and experimental shortcomings. The theoretical drawback is the lack of success in building a bootstrap theory that will lead to a unique solution. The experimental problem is the observation of long-range correlations. The two problems are presumably related and it may be anticipated that a pure multiperipheral model cannot provide the correct description of high energy scattering. The prime question - what is then the correct theory - is left unanswered. There are several different approaches that one may choose to try to find the answer. One is to assume that a multiperipheral chain generates a moving Pomeron pole at $\alpha_P(o)=1$ and upon iteration leads to all the cuts discussed in chapters 13 and 14. Another is to say that the multiperipheral chain builds up a moving pole with $\alpha_P(o)>1$ which is to be identified with the first term in an eikonal expansion, in the spirit of chapter 11. Other alternative approaches were mentioned in chapter 15. All these models have a multiperipheral ingredient but the way this ingredient is exploited differs from one to the other.

One of the models discussed in chapter 15 included a multiperipheral production mode that built a constant cross section and a diffractive mode that led to a logarithmically increasing cross section. Actually, as far as phenomenology is concerned, we may use a division of σ_I into diffractive and nondiffractive parts and ask whether the latter can be identified with a multiperipheral component. A key question is indeed the energy behavior. If σ_I increases logarithmically with s we can ask what amount of this increase is contributed by the diffractive part - the latter is identified as that part of the fragmentation distribution $\rho^C_{AB}(s,t,M^2)$ which does not decrease with s. Future experimental investigation will provide the answer. In chapter 6 we saw that unless the triple Pomeron coupling vanishes when all three trajectories are at $\alpha=1$, it will contribute a $\ln\ln s$ behavior to σ_I for a moving Pomeron, and a $\ln s$ for a fixed Pomeron. Since for practical purposes the Pomeron resembles a fixed singularity it may very well contribute a logarithmically increasing cross section. However, the nondiffractive part can also participate in the increasing cross section. In particular, if the central plateau increases logarithmically so will the cross section.

Paradoxical as it may seem, it can be difficult to distinguish between a logarithmically increasing and a logarithmically decreasing correction. Thus the first order in $\ln s$ in both $-(1+\frac{2\alpha'}{b}\ln s)^{-1}$ and $\ln(1+\frac{2\alpha'}{b}\ln s)$ have the same $\frac{2\alpha'}{b}\ln s$ form. The first expression is characteristic of a negative cut, the second of a triple Pomeron contribution, where α' is the slope of the moving pole and b is the slope parameter of $|\beta(t)|^2 = e^{bt}$. Since $\frac{2\alpha'}{b}$ may be very small (e.g., <0.05) the first order in $\ln s$ will be a good approximation for present s values. Moreover, if that is not enough, we cannot escape from the fact that numerically it is hard to distinguish between a power and a $\ln s$ behavior.[33)] Thus, over the whole s range of 100 to 600 GeV^2, one finds that $\ln s = 1.97\ s^{0.185}$ to within 1%. Hence, the interpretation of experimental results will still be ambiguous for some time to come.

The multiplicity distribution can be accounted for in a hybrid model which has a short-range (multiperipheral) component σ_S and a diffractive component σ_D:

$$\sigma_I = \sigma_S + \sigma_D. \tag{20.1}$$

Such a model was described in chapter 3 where a simple calculation showed us that the data lead to a ratio of $\sigma_S:\sigma_D \approx 4:1$ at NAL energies. Such a hybrid model was suggested a long time ago by Wilson.[34)] In this picture one assumes that the diffractive component has a bounded σ_n distribution, falling rapidly with n. A clear test of this hybrid model would be the observation of a dip or a gap in the prong distribution. This should occur at energies high enough so that the multiperipheral distribution separates from the diffractive one. Unfortunately, this has not yet been observed.

The separation of the distribution into two parts means that the generating function can be split in the form

$$Q(s,z) = \frac{\sigma_D}{\sigma_I} Q_D + \frac{\sigma_S}{\sigma_I} Q_S \tag{20.2}$$

where Q_D and Q_S are the normalized $(Q(z=1)=1)$ generating functions for the diffractive and short-range components respectively. If the diffractive distribution contains only finite values of n independent of s,[*)] and if Q_S has the familiar short-

*Note that this is not the case in the PPP distribution discussed in chapter 18 or in the model of eq.(15.21) where a long diffractive tail existed up to $n \sim \ell n\, s$.

range form, we may write

$$Q_D = Q_D(z) \qquad Q_S = q(z)s^{r(z)} \tag{20.3}$$

all subject to the condition $Q(z=1)=1$. In a multiperipheral model with fixed pole (α_R) input one finds

$$r(z) = (2-2\alpha_R)(z-1). \tag{20.4}$$

In general, we know that $r(z)$ has to pass through zero at $z=1$. Hence Q is dominated by Q_S for $z>1$ and Q_D for $z<1$. In particular,

$$p(z) = \lim_{s\to\infty} \frac{\ln Q}{\ln s} = \begin{cases} r(z) & z>1 \\ 0 & z<1 \end{cases} . \tag{20.5}$$

It turns out that this discontinuous behavior has an analogue in thermodynamical terminology. To understand this let us start with the multiperipheral component. This has been called the Feynman gas by Wilson.[35)] Let us think of a one-dimensional gas of particles which have only short-range forces between them. The dimension is rapidity and the forces lead to short-range correlations. This analogy can be carried further by noting that $Q(z,Y)$ and $\sigma_n(Y)$ can be considered as grand-canonical and microcanonical partition functions respectively.[36)] $\sigma_n(Y)$ has indeed a fixed number of particles and one overall variable $Y\sim \ln s$. Y represents the volume of the gas and since the dependence on it is linear we find $\partial \ln Q/\partial Y = \ln Q/Y$ which is naturally identified with pressure. Eq. (20.5) is then interpreted as a phase transition at $z=1$ from a state with positive pressure at $z>1$ to a state with zero pressure at $z<1$. Zero pressure means that the fluid clings to the container walls - a picturesque terminology to describe the leading role of diffractive dissociation.

The generating function of (20.2) is supposed to contain the information about the partial cross sections and their energy dependence. We have discussed its behavior near $z=1$, the point used for calculating correlations. Can we continue it down to $z=0$ to deduce all partial cross sections? Probably not. The reason is that if there exist multiperipheral and diffractive amplitudes there must also be some interference between them. Then we should really write, in place of (20.2),

$$Q = \frac{\sigma_D}{\sigma_I} Q_D + \frac{\sigma_S}{\sigma_I} Q_S + \frac{\sigma_{SD}}{\sigma_I} Q_{SD} \; . \tag{20.6}$$

However, if $\sigma_{n,D}$ extend over only a finite range in n, and $\sigma_{n,S}$ are falling like some power $s^{2\alpha_R-2}$, we may expect $\sigma_{n,SD}$ as well as σ_{SD} to behave like s^{α_R-1}. Q_{SD} will be function of z only, reflecting the finite range in n of the diffractive component. Using any $\alpha_R<1$ we see that near $z=1$ the interference term vanishes for large s, leading us back to (20.2). However, it may overtake the multiperipheral term at some point $0<z<1$, at which $\alpha_R-1=r(z)$.

With the particular choice of eq. (20.4) this occurs at $z=\frac{1}{2}$.

The striking behavior of eq. (20.5) should, in principle, provide a clear test of this type of hybrid model. Unfortunately, even with the very high energies now available, we are still not in a position to observe this effect.[33] To see this, let us use the definition $Q=\frac{1}{\sigma_I}\sum_{n=2}^{\infty} z^{n-2}\sigma_n$ for the sum over n charged particles. At z=0 one then obtains

$$\frac{\ln Q}{\ln s} = \frac{\ln (\sigma_2/\sigma_I)}{\ln s}$$

which, in the NAL energy range, has numerical values around -0.4. Moreover, the two-prong inelastic cross section σ_2 is still decreasing, therefore, going to even larger s, it is not clear if we are going to approach zero, as obviously necessitated by a finite asymptotic ratio of σ_2/σ_I. Therefore, only if this ratio is finite, and when $\langle n\rangle \gg 2$, do we have a chance to observe the behavior indicated in (20.5).

Although the phase transition of the fluid from a gas to a liquid may be a somewhat exaggerated description when one deals with the small numbers of observed secondaries in high-energy collisions, the direct use of the generating function Q(s,z) in constructing hybrid models may be a constructive step. In part IV we discussed multiperipheral models using the s-channel unitarity equation. We have already pointed out in chapter 13 that these are actually equations for Q(s,z). In building hybrid models one tries to find a way to correct the solution of the oversimplified model, which may indeed be analogous to introducing corrections to an ideal gas picture. In dealing directly with Q(s,z) one has the comfort - and the challenge - of using an experimentally measurable object. Recent work has suggested adding long-range forces between the constituents of the liquid either for phenomenological fits[37)] or to represent the effect of Pomeron cuts.[38)] We may confidently expect that more research in this direction is still to come.

REFERENCES

1. T.T. Chou and C.N. Yang, Phys. Rev. 170, 1591 (1968); Phys. Rev. Letters 20, 1213 (1968); Phys. Rev. 175, 1832 (1968).
2. L. Durand III and R. Lipes, Phys. Rev. Letters 20, 637 (1968).
3. J.D. Bjorken, invited talk at the Int. Conf. on duality and symmetry in hadron physics, Tel—Aviv (1971) (ed. E. Gotsman, Weizmann Press, Jerusalem), p. 98.
4. M. Bishari, D. Horn and S. Nussinov, Nucl. Phys. B36, 109 (1972).
5. L. Van Hove, Rev. Mod. Phys. 36, 655 (1964).
6. R.P. Feynman, "Photon—Hadron Interactions," (Benjamin, Mass. 1972).
7. A. Bialas and Th. W. Ruijgrok, Nuovo Cim. 39, 1061 (1965).
8. D. Horn, Phys. Reports 4C, 1 (1972).
9. S. Feinberg and J. Grunhaus, Nucl. Phys. B43, 147 (1972).

10. M. Kugler and R.G. Roberts, preprint.
11. J. Bartke, Lectures at Herceg-Novi School (1970).
12. D. Sivers and G.H. Thomas, Phys. Rev. D6, 1961 (1972).
13. F. Cerulus, Nuovo Cim. 19, 528 (1961).
J. Shapiro, Suppl. Nuovo Cim. 18, 40 (1960).
14. D. Horn and R. Silver, Phys. Rev. D2, 2082 (1970); Ann. of Phys. 66, 509 (1971).
15. G.H. Thomas, Phys. Rev. D7, 2058 (1973).
16. L. Van Hove, Nuovo Cim. 28, 708 (1963).
17. F. Lurcat and P. Mazur, Nuovo Cim. 31, 140 (1964).
18. E. Fermi, Progr. Theor. Phys. (Kyoto) 5, 570 (1950).
19. R. Hagedorn, Nuovo Cim. 52A, 1336 (1967); Nuovo Cim. Suppl. 6, 311 (1968).
20. R. Hagedorn, and J. Ranft, Nuovo Cim. Suppl. 6, 169 (1968).
21. S. Frautschi, Phys. Rev. D3, 2821 (1971).
22. A. Krzywicki, Phys. Rev. 187, 1964 (1969).
S. Fubini and G. Veneziano, Nuovo Cim. 64A, 811 (1969).
K. Bardakci and S. Mandelstam, Phys. Rev. 184, 1640 (1969).
S. Fubini, D. Gordon and G. Veneziano, Phys. Letters 29B, 679 (1969).
23. J. Erwin, W. Ko, R.L. Lander, D.E. Pellett, and P.M. Yaeger, Phys. Rev. Letters 27, 1534 (1971).
24. P. Carruthers and M. Duong-Van, Phys. Letters 41B, 597 (1972).
25. R.C. Hwa, Phys. Rev. D1, 1790 (1970); Phys. Rev. Letters 26, 1143 (1971).
26. M. Jacob and R. Slansky, Phys. Letters 37B, 408 (1971); Phys. Rev. D5, 1847 (1972).
27. K. Gottfried and O. Kofoed-Hansen, Phys. Letters 41B, 195 (1972).
28. H.D.I. Abarbanel and G.L. Kane, Phys. Rev. Letters 30, 67 (1973).
29. M. Jacob, Rapporteur talk at Chicago Conf. 1972, Vol. 3, p. 373.
30. E.L. Berger, D. Horn and G.H. Thomas, Phys. Rev. D7, 1412 (1973).
31. D. Horn and A. Schwimmer, Nucl. Phys. B52, 627 (1973).

32. L. Caneschi and A. Schwimmer, Phys. Letters 33B, 577 (1970).
33. M. Bander, preprint NAL-THY-98, 1972.
34. K. Wilson, Acta Phys. Austr. 17, 37 (1963).
35. K. Wilson, Cornell preprint, CLNS-131 (1971).
36. A.H. Mueller, Phys. Rev. D4, 150 (1971).
    R.C. Arnold, Phys. Rev. D5, 1724 (1972), Preprint ANL/HEP 7241 (1972).
    M. Bander, Phys. Rev. D6, 164 (1972).
37. R.C. Arnold and G.H. Thomas, preprint ANL/HEP 7257 (1972).
38. T.L. Neff, Phys. Letters 43B, 391 (1973).

Summary and Outlook

We should like to close with a brief summary of where our theoretical understanding of strong interactions at very high energies now stands.

First of all we have a very useful qualitative - or even semiquantitative - guide in the simple extension of the two-body Regge pole phenomenology to high energy production processes; that is, in the Mueller analysis and in the MUREX hypothesis. But we must be careful to not carry this simple picture too far. It is experimentally clear that long-range correlations exist, and the Regge pole picture is explicitly one containing short-range correlations only. Long-range correlations are probably related to the existence of j-plane branch points in addition to or instead of simple Regge poles. Therefore, one plausible direction for future development is to try to modify or replace the Regge pole phenomenology with Regge cuts and to study how far these can go toward resolving the failures of the pure pole picture, without, of course, distorting the success of that picture.

We have also described various models which, in varying degrees, purport to provide a more fundamental theoretical basis for our understanding of high energies. These include multiperipheral models, droplet models, nova or fireball models and field theories. There are, in addition, models dealing with hadronic structure which we have not mentioned, such as the parton model and dual models. The parton model has been excluded because in its present form it leads to only qualitative results which are not different from the Regge Mueller approach. While it is true that thinking in terms of partons motivated Feynman to suggest the existence of scaling, it seems that the Regge pole picture is a simpler generalization from two-body scattering amplitudes and is based on familiar techniques. Dual models have been excluded because the planar dual diagrams do not represent Pomeron effects. They were useful in exploring properties of six- and eight-point functions and checking their expected Regge pole forms. However, in order to relate to diffractive phenomena one has to go into complicated, and unsettled, structures of loop diagrams.

Among the models we have included, the multiperipheral model stands out both because it leads naturally to the at least semisuccessful Regge phenomenology, and because of its simplicity. Nevertheless, it cannot be strictly correct, and it is clear that modifications which try to combine the main features of multiperipheralism with something introducing long-range correlations will be of great interest. It may be that such modifications should be in terms of models of the hybrid or two component type, or it may be that they will involve changes in the multiperipheral matrix element itself, such as are provided for example by introducing absorption. The modified models may continue to be most naturally expressed in the t-channel j-plane language, as is the multiperipheral model, or if the j-plane structure becomes too complicated, it may be that some other language, such as the thermodynamical analogy, is to be preferred. Whatever the correct approach, it is likely to be built on, but will certainly not be identical with, the multiperipheral model.

The use of field theory as a guide is also an approach worth keeping in mind, but caution is required about how to connect a field theory to hadron physics. The relevance must doubtless be through the quark-gluon model, but if it is, then detailed predictions will require an understanding of the bound state structure of individual hadrons in terms of quarks. Nevertheless, more qualitative suggestions from field theory, such as the statement that the Froissart bound is saturated, may well be transferable to strong interactions as they stand.

In any event, we can confidently expect continued rapid progress in this field on the experimental end at least, and perhaps theorists will be stimulated to match this with a steadily improving theoretical understanding as well.

APPENDICES

A

The Optical Theorem

The conventional optical theorem follows immediately from the definition of the total cross section and s-channel unitarity for the elastic scattering amplitude. We recall that, for large s, the cross section for A+B→anything is

$$\sigma_T(s) = \frac{1}{2s}\sum_{n=2}^{\infty}\int d\Phi_n \left| T_{2\to n}(p_1 p_2 \to q_1 \cdot\cdot q_n)\right|^2 \qquad (A1)$$

while unitarity for the elastic amplitude reads

$$2A(s,t) = \sum_{n=2}^{\infty}\int d\Phi_n T_{2\to n}(p_1 p_2 \to q_1 \cdot\cdot q_n)\cdot$$

$$\cdot\; T_{2\to n}^{*}(p_1' p_2' \to q_1 \cdot\cdot q_n). \qquad (A2)$$

Thus it follows that

$$A(s,o) = s\,\sigma_T(s). \qquad (A3)$$

The generalizations of this relationship to a connection between more complicated inclusive cross sections and forward multibody amplitudes is more subtle.[1)] Let us look first at the one particle inclusive cross section for A+B→C+anything. This is given by

$$\frac{d\sigma}{d^3\vec{q}} = \frac{1}{2s} \sum_n \int d\Phi_n \left[\sum_{i=1}^{n} \delta^3(\vec{q}-\vec{q}_i)\right] \cdot$$

$$\cdot \left|T_{2\to n}(p_1 p_2\ q_1 \cdot\cdot q_n)\right|^2 \tag{A4}$$

$\omega\, d\sigma/d^3q$ is a Lorentz invariant function of $s=(p_1+p_2)^2$, $M^2=(p_1-q+p_2)^2$, and $t=(p_1-q)^2$. Now, with our normalizations,

$$\frac{T_{2\to n}(p_1 p_2 \to q_1 \cdot\cdot q_{i-1} q\ q_{i+1} \cdot\cdot q_n)}{\sqrt{4E_1 E_2}\ \sqrt{2\omega_1\ 2\omega_2 \cdots 2\omega_n}}$$

$$= \frac{1}{\sqrt{2\omega}} \langle q_1 \cdot\cdot q_{i-1}\ q_{i+1} \cdot\cdot q_n{}^{(-)} | j_c(o) | p_1 p_2{}^{(+)} \rangle \tag{A5}$$

where (for the case where c is a boson described by a renormalized Heisenberg field Φ_c) we have

$$(\Box + m_c{}^2)\ \Phi_c(x) = j_c(x). \tag{A6}$$

An analogous definition of j_c obtains for fermions. Thus we have (setting n-1→n)

$$\frac{d\sigma}{d^3q} = \frac{1}{2s}\frac{(4E_1E_2)}{2\omega}\left(\frac{1}{2\pi}\right)^3 \sum_n \int \frac{d^3q_1}{(2\pi)^3} \cdots \frac{d^3q_n}{(2\pi)^3} \cdot$$

$$\cdot\ (2\pi)^4 \delta^{(4)}(P - q - \sum_i q_i) \cdot$$

$$\cdot |\langle q_1 \cdots q_n^{(-)} | j_c(o) | p_1 p_2^{(+)} \rangle|^2$$

$$= \frac{1}{2s}\frac{4E_1E_2}{2\omega}\left(\frac{1}{2\pi}\right)^3 \int d^4x e^{-iqx} .$$

$$\cdot\ \langle p_1 p_2^{(+)} | j_c(x) j_c(o) | p_1 p_2^{(+)} \rangle . \qquad \text{(A7)}$$

On the other hand we can also consider the forward process $A+\overline{C}+B \to A+\overline{C}+B$ with momenta p_1, $-q$, p_2 respectively. The T-matrix for this is

$$T_{A\overline{C}B\to A\overline{C}B} = i(4E_1E_2)\int d^4x e^{-iqx} .$$

$$\cdot \langle p_1 p_2^{(-)} | (j_c(x) j_c(o))_+ | p_1 p_2^{(+)} \rangle \qquad \text{(A8)}$$

and the discontinuity of the T-matrix in the energy $(p_1 - q + p_2)^2 = M^2$ is

$$A_{A\overline{C}B\to A\overline{C}B}(p_1, -q, p_2 \to p_1, -q, p_2)$$

$$= \frac{1}{2}(4E_1E_2)\int d^4x e^{-iqx} \langle p_1 p_2^{(-)} | \left[j_c(x), \right.$$

$$\left. , j_c(o) \right] | p_1 p_2^{(+)} \rangle . \qquad \text{(A9)}$$

A is evidently Lorentz invariant, and depends on an "energy" $M^2=(p_1-q+p_2)^2$ and two independent "subenergies" $s=(p_1+p_2)^2$ and $t=(p_1-q)^2$. The third subenergy, $u=(p_2-q)^2$, is not independent.

If we could ignore the fact that the true absorptive part has incoming boundary conditions, and other subtleties regarding which sheets of the variables s, t and M^2 $\omega d\sigma/d^3q$ and A are to be evaluated on, then comparing (A9) with (A7) yields[1)]

$$\omega_C \frac{d\sigma_{AB}^{\;C}}{d^3\vec{q}_C} = \frac{1}{2s}\left(\frac{1}{2\pi}\right)^3 A_{A\overline{C}B\to A\overline{C}B}(p_1-qp_2;p_1-qp_2). \tag{A10}$$

Now in actual fact we cannot ignore the boundary conditions and sheet structure; therefore, the A which actually must stand on the right-hand side of eq. (A9) is not the absorptive part of a physical forward 3→3 amplitude. It is rather associated with an unphysical 3→3 amplitude obtained by analytic continuations from the physical amplitude.[2)] Let us outline how this relationship is obtained.

The first point is that the inclusive cross section is the absolute square of a matrix element; that is to say, the product of one matrix element evaluated at energy s just above the cut in s times the same matrix element evaluated with s just below its cut. Hence if we think of the forward 3→3 amplitude as a function of both an initial s and a final s, call them s_1 and s_2, we are interested in $T(s_1,M^2,s_2)|_{s_{1,2}=s\pm i\varepsilon}$, where s is real, for comparison with the inclusive cross section. The physical forward 3→3 amplitude, in contrast, is $T(s_1,M^2,s_2)|_{s_1=s_2=s+i\varepsilon}$.

Next, the object in eq. (A10) is the absorptive part in M^2 of T; that is, we want

$$2iA = T(+,+,-) - T(+,-,-) \tag{A11}$$

in an obvious notation. The physical 3→3 absorptive part is not this, but rather

$$2iA = T(+,+,+) - T(+,-,+). \tag{A12}$$

Hence to obtain the absorptive part to be used in the generalized optical theorem we must continue the physical 3→3 absorptive part in s_2 from $s_2=s+i\varepsilon$ to the point $s_2=s-i\varepsilon$.

The final point to be made is that the 3→3 absorptive part includes disconnected graphs, as well as semidisconnected ones (in which both C lines connect to only one AB blob). Obviously, these are not contained in the inclusive cross section; hence, the A used in eq. (A10) must be only the analytic continuation of the completely connected part of the $3 \to 3$ amplitude.

With these modifications, eq. (A10) is valid.[2] Its usefulness lies not in the direct comparison of experimentally measured quantities on both sides of the equation, but as a theoretical tool, because of the fact that a Regge pole expansion could be just as valid for T(+,+,-) as it is for any other physical scattering amplitude. Hence it makes plausible the use of Regge phenomenology for inclusive cross sections in the manner described in chapter 6.

It is clear that generalizations of (A10) to two, three, and more body inclusive processes are also possible. Their derivation we leave to the reader.

B

The Froissart Bound

One of the most important of the few general and model independent theoretical statements which can be made about very high energies is the Froissart bound.[3)] This bound states that the total cross section for any process cannot grow faster than $(\ell n\ s)^2$ as $s\to\infty$:

$$\sigma_T(s) \leq \text{const } (\ell n\ s)^2. \tag{B1}$$

An unsophisticated proof of this is given below, for the simple case of the scattering of two spinless particles.[4)]

Let $T(s,t)$ be the elastic scattering amplitude, and write the s-channel partial wave expansion:

$$T(s,t) = \sum_{j=o}^{\infty} (2j+1)\ P_j(x_s)\ T(s,j) \tag{B2}$$

where $x_s=1+t/2q_s^2$ and $q_s^2=s/4-\mu^2$. The inverse of (B2) is

$$T(s,j) = \frac{1}{2}\int_{-1}^{1} dx_s\ P_j(x_s)\ T(s,t) \tag{B3}$$

and with our normalization we have

$$T(s,j) = 16\pi\ \frac{\eta_j(s)\ e^{2i\delta_j(s)} - 1}{2i} \tag{B4}$$

in terms of the absorption and the phase shift. We note, from (B4), that

$$|T(s,j)| \leq 16\pi. \tag{B5}$$

The s-channel absorptive part may, from (B2), be written

$$A(s,t) = \sum_{j=o}^{\infty} (2j+1)\ P_j(x_s)\ \mathrm{Im}\ T(s,j) \tag{B6}$$

and we note that $\mathrm{Im}\ T(s,j)=8\pi(1-\eta\cos 2\delta)$ so that

$$16\pi \geq \mathrm{Im}\ T(s,j) \geq \frac{1}{16\pi}\ |T(s,j)|^2 \geq 0. \tag{B7}$$

Now (B2), as written, holds in the physical region $-1\leq x_s\leq 1$. It can be extended to hold for complex x_s as well, because of the analytic structure of T(s,t). The expansion is valid inside the

Lehmann ellipse, an ellipse in the x_s plane with foci at ±1 passing through $\pm(1+t_o(s)/2q_s^2)$ where $t_o(s)$ is the position of the lowest singularity in t of T(s,t).*)

Let's assume that T(s,t) has a polynomial bound in s in the Lehmann ellipse:

$$A(s,t) \le |T(s,t)| < s^N \qquad \text{(B8)}$$

for some N, as $s\to\infty$. Then at the edge of the ellipse, (B6) says that

$$A(s,t) = \sum_j (2j+1)P_j\left(1+\frac{t_o(s)}{2q_s^2}\right) \text{Im } T(s,j). \qquad \text{(B9)}$$

But since $P_j(x)>0$ if $x>1$, and since Im $T(s,j)>0$, this is a sum of positive terms. Thus each term is bounded by (B8); consequently

$$0 \le \text{Im } T(s,j) \le \frac{s^N}{(2j+1)\ P_j\left(1+\frac{t_o(s)}{2q_s^2}\right)} \qquad \text{(B10)}$$

for each j. Using some simple properties of P_j, namely that

---

* Coulomb scattering produces an infinite total cross section and thereby violates the Froissart bound. The reason for this is that due to the zero mass of the photon $t_o(s)=0$ so that the expansion cannot be extended outside of $-1\le x_s\le 1$.

$$P_j(x) > \frac{c_o}{2j+1}\left(1 + \sqrt{2(x-1)}\right)^j, \tag{B11}$$

(B10) yields, together with (B7), that

$$\begin{aligned} 0 &\le \frac{1}{16\pi}\,|T(s,j)|^2 \le \operatorname{Im} T(s,j) \\ &\le \frac{s^N}{c_o}\left(\frac{1}{1+\sqrt{\dfrac{t_o(s)}{q_s^{\,2}}}}\right)^j . \end{aligned} \tag{B12}$$

Now, when $s \to \infty$, $t_o(s)/q_s^{\,2} \to 0$. So we obtain

$$\begin{aligned} 0 &\le \frac{1}{16\pi}\,|T(s,j)|^2 \le \operatorname{Im} T(s,j) \\ &\le s^N \exp\left(-j\sqrt{\frac{t_o(s)}{q_s^{\,2}}}\right). \end{aligned} \tag{B13}$$

Hence in the partial wave sum we can neglect terms with $j > j_{max}$, where

$$j_{max} = N\sqrt{\frac{q_s^{\,2}}{t_o(s)}}\;\ell n\, s. \tag{B14}$$

Thus, returning to the physical region, we have from (B2)

$$A(s,t) \le |T(s,t)| \le \sum_j (2j+1)\,|P_j(x_s)|\,|T(s,j)| \tag{B15}$$

$$\text{and} \quad A(s,o) \leq \sum_{j=o}^{j_{max}} (2j+1)|T(s,j)|$$

$$\leq 16\pi \sum_{j=o}^{j_{max}} (2j+1) \sim j_{max}^{2} \qquad \text{(B16)}$$

and finally

$$\sigma_T(s) \leq \text{const.} \frac{q_s^{2} \ell n^2 s}{s} \sim \ell n^2 s. \qquad \text{(B17)}$$

To summarize, this result required the following assumptions:

(i) Analyticity in t, permitting the use of the partial wave expansion inside the Lehmann ellipse, and

(ii) Polynomial boundedness in s.

The physical implications of all of this, and in particular of eq. (B14), is that in an impact parameter description the scatterer cannot be bigger than a black disk of radius proportional to ℓn s. This means that scattering which saturates the Froissart bound is described by an amplitude

$$T(s,t) = 8\pi i s \int_{o}^{R_o \ell n\, s} b db\, J_o(b\sqrt{-t})$$

$$= 4\pi i s R_o \ell n\, s\, J_1(R_o\sqrt{-t}\, \ell n\, s)/\sqrt{-t}, \text{(B18)}$$

which leads to a total cross section

$$\sigma_T = 2\pi R_o^2 \ln^2 s \tag{B19}$$

and very strong ($\ln^2 s$) shrinkage of the forward peak. We also find that $\sigma_{e\ell}/\sigma_T = 1/2$.

In the j-plane (see appendix D) the amplitude is

$$T(t,j) \propto \left((j-1)^2 - R_o^2 t\right)^{-\frac{3}{2}} \tag{B20}$$

so there are, for $t<0$, two complex conjugate branch points of the $(j-\alpha_c)^{-3/2}$ type at

$$\alpha_c = 1 \pm iR_o \sqrt{-t}. \tag{B21}$$

These coalesce, at $t=0$, into a third order pole at $j=1$; hence the $\ln^2 s$ behavior of the cross section.

Saturation of the Froissart bound yields, as we have seen, $\sigma_{e\ell} = 1/2\ \sigma_T$. More generally, we must, of course, require

$$\sigma_{e\ell}(s) \leq \sigma_T(s), \tag{B22}$$

and this also yields constraints on the high energy behavior of diffraction peaks. If we couple (B22) with the optical theorem (appendix A) we find the inequality

$$\frac{\sigma_T^2(s)}{16\pi b(s)} \leq \sigma_{e\ell}(s) \tag{B23}$$

where we have assumed that to a sufficiently good approximation we can write

$$(d\sigma/dt)/(d\sigma/dt)_{t=0} = e^{b(s)t}. \tag{B24}$$

If ReT(s,o)/A(s,o)=0, the inequality in (B23) becomes an equality.

The inequalities (B22) and (B23) tell us that

$$\sigma_T(s) \leq 16\pi b(s). \tag{B25}$$

Thus we cannot have a growing total cross section without also having shrinkage of the diffraction peak.

C

The Pomeranchuk Theorem

Another important consequence of what is considered by most people to be basic and unimpeachable principles alone is the following theorem: If the total cross section behaves asymptotically like a power of $\ln s$, then either particle and antiparticle cross sections become asymptotically equal: $\sigma_{AB}(s)\to\sigma_{A\bar{B}}(s)$, or else the forward amplitude becomes dominantly real asymptotically: $\mathrm{Re}T(s,o)/A(s,o)\to\infty$. What is usually referred to as the Pomeranchuk Theorem[5)] is the first of these two alternatives.

The theorem follows from the assumption of s-channel analyticity and s-u crossing symmetry. Let us begin with the consequences of analyticity. We believe that the scattering amplitude $T(s,t)$ for an elastic scattering process $A+B\to A+B$ is an analytic function of s for fixed t, and that we can write

$$T(s,t) = a(t)+b(t)s + \frac{s^2}{\pi}\int_{s_o}^{\infty} \frac{A(s',t)}{s'^2(s'-s)}\,ds' + \frac{u^2}{\pi}\int_{u_o}^{\infty} \frac{\bar{A}(u',t)}{u'^2(u'-u)}\,du'. \tag{C1}$$

We have made two subtractions to allow for the experimental fact that $|T| \sim s$ at large s; according to the experiments, and to the Froissart bound, it seems that more than two subtractions are unnecessary.

Because of the optical theorem, we know that on the right-hand cut, for s above threshold, the absorptive part $A(s,t)$ of the amplitude satisfies

$$A(s,o) = s\ \sigma_{AB}(s), \tag{C2}$$

where $\sigma_{AB}(s)$ is the total cross section for A on B.

Because of crossing, interchanging s and u is equivalent to interchanging the reactions $A+B\to A+B$ and $A+\bar{B}\to A+\bar{B}$; thus on the left-hand cut, for $u>u_o$, we have for the absorptive part

$$A(u,o) = u\ \sigma_{A\bar{B}}(u), \tag{C3}$$

where $\sigma_{A\bar{B}}(u)$ is the total cross section for A on $\bar{B}$. More generally, for $t\neq 0$ and s above threshold,

$A(s,t)$ is the imaginary part of the $A+B\to A+B$ amplitude while $\overline{A}(u,t)$ is the imaginary part of the $A+\overline{B}\to A+\overline{B}$ amplitude for u above threshold.

Now, experimentally, it is suggested that $\sigma_{AB}(s)\to$const. as $s\to\infty$, as does $\sigma_{A\overline{B}}(u)$ as $u\to\infty$. Let us for the moment believe this to be true. The dispersion integrals, at $t=0$, then behave like $s \ln s$ and $u \ln u$, for the right-hand cut and left-hand cut terms respectively, at large s and large u. Then as $s\to\infty$, at $t=0$ we have

$$T(s,o) \to a+bs-\frac{\sigma_{AB}}{\pi}s \ln(-s)-\frac{\sigma_{A\overline{B}}}{\pi}u \ln(-u). \quad (C4)$$

But $u\to -s$ in this limit, so that in fact

$$T(s,o) \to a+bs+is\sigma_{AB}-s\frac{(\sigma_{AB}-\sigma_{A\overline{B}})}{\pi}\ln s. \quad (C5)$$

Hence if $\mathrm{Re}T(s,o)/A(s,o)\sim o(\ln s)$, we must conclude that $\sigma_{AB}=\sigma_{A\overline{B}}$ -- i.e. that particle and antiparticle total cross sections approach each other. This is the simplest form of the "Pomeranchuk theorem."

We may note that if $\sigma_{AB}\neq\sigma_{A\overline{B}}$, so that $\mathrm{Re}T(s,o) \sim s \ln s$ as $s\to\infty$, then the forward peak in elastic scattering must shrink at least like $\ln^2 s$, that is, the slope parameter $b(s)$ must grow at least as fast as $\ln^2 s$ to prevent $\sigma_{e\ell}$ from growing with s, and thus becoming bigger than the total cross section itself.

We have mentioned that some theoretical

models as well as some experiments suggest that $\sigma_T \not\to$ const., but rather behaves like $1/(\ln s)^{\alpha}$, or like $s^{-\varepsilon}$ or even like $(\ln s)^2$ as $s \to \infty$. Experiment does not unequivocally rule out these possibilities so it may be worth while to mention their consequences for the Pomeranchuk theorem.

Firstly, if σ behaves like a power of s (a falling power, naturally, to avoid violation of the Froissart bound), then A(s,t)/s vanishes asymptotically. From eq. (C1) we then conclude that ReT(s,t)/s vanishes in the same way as does A(s,t)/s, so that ReT/A→const., whether or not particle and antiparticle cross sections become equal.

Secondly, if σ increases with a power of $\ln$ s, then one can show[17)] from unitarity that ReT/A $\ln s \to 0$ and the Pomeranchuk theorem follows without further assumption. If σ decreases with a power of $\ln$ s, then, as in the constant cross section case, either ReT/A $\sim \ln s$ or the Pomeranchuk theorem holds.

We may also look at the behavior of T(s,t) for large s with $t \neq 0$, using eq. (C1). If for $t \neq 0$ it is true that ReT/A→0, then we see that A(s,t) $\to \bar{A}(s,t)$ as $s \to \infty$ for all t. Thus we must expect forward peaks for particle and antiparticle cross sections to become equal.

D

The j-plane

In the text much use has been made of the equivalence of the s,t and t,j descriptions of high energy two-body processes. The asymptotic behavior in energy s at a fixed momentum transfer t of a scattering amplitude T(s,t) is related to the singularities in angular momentum j at fixed energy t of a partial wave amplitude T(t,j). We wish to construct, and to elaborate on, this relationship here.[6)]

Let us suppose that T(s,t) is the amplitude for two-body elastic scattering $A+A\rightarrow A+A$. Because of $s\leftrightarrow t$ crossing, we are invited to assume that this same function, when continued from the region $s>4m_A{}^2$, $t<0$ (the s-channel physical region) to the region $t>4m_A{}^2$, $s<0$ (the t-channel physical region), is also the amplitude for the reaction $A+\overline{A}\rightarrow A+\overline{A}$.

We begin, as in the appendix on the Pomeranchuk theorem, with the dispersion relation in s at fixed t which we are accustomed to believe the amplitude to satisfy. As we have remarked elsewhere, two subtractions are indicated experimentally to suffice. Thus

$$T(s,t) = a(t)+b(t)s+\frac{s^2}{\pi}\int_{s_o}^{\infty}\frac{A(s',t)ds'}{s'^2(s'-s)}$$

$$+\frac{u^2}{\pi}\int_{s_o}^{\infty}\frac{A(u',t)du'}{u'^2(u'-u)}\,. \tag{D1}$$

In the region $t>4m_A^2$, $s<0$, where $T(s,t)$ describes the reaction $A+\overline{A}\to A+\overline{A}$, in the center of mass system $E=\sqrt{t}/2$ is the energy of each particle while $x_t=1+s/2q_t^2=-1-u/2q_t^2$ is the cosine of the t-channel c.m. scattering angle, where $q_t=\sqrt{t/4-m_A^2}$ is the c.m. momentum.

When viewing $T(s,t)$ as an amplitude in which t is the energy and s is the momentum transfer it is convenient to break T up into parts even and odd in x_t. We therefore write

$$T(s,t) = \frac{1}{2}\left[T^+(s,t)+T^+(u,t)\right]$$

$$+\frac{1}{2}\left[T^-(s,t)-T^-(u,t)\right].$$

The dispersion relation (D1) can now be written as dispersion relations in x_t for fixed t for T^+ and T^-. These functions are chosen to have only right-hand cuts, which are determined by (D1). We write

$$T^{+}(s,t) = \bar{a}(t) + \frac{x_t^2}{\pi}\int_{x_o}^{\infty}\frac{A^{+}(t,z)dz}{z^2(z-x_t)} \tag{D2}$$

and $$T^{-}(s,t) = \bar{b}(t)x_t + \frac{x_t^2}{\pi}\int_{x_o}^{\infty}\frac{A^{-}(t,z)dz}{z^2(z-x_t)} \tag{D3}$$

where $\bar{a}$ and $\bar{b}$ are related to a and b in an obvious way, and where

$$A(s',t) = \frac{1}{2}\left[A^{+}(t,z)+A^{-}(t,z)\right]$$

$$\bar{A}(u',t) = \frac{1}{2}\left[A^{+}(t,z)-A^{-}(t,z)\right]$$

with $s'=-2q_t^2(1-z)$ and $u'=-2q_t^2(1+z)$. Note that $x_o=1+s_o/2q_t^2$; thus for $q_t^2>0$, $x_o>1$.

Let us now decompose the amplitudes into t-channel partial waves, with the usual definition

$$T^{\pm}(s,t) = \sum_{j\,\substack{\text{even}\\ \text{odd}}} (2j+1)\; P_j(x_t)\; T^{\pm}(t,j). \tag{D4}$$

The dispersion relations (D2) and (D3) together with the magic identity[7)]

$$\frac{1}{z-x_t} = \sum_j (2j+1)\; P_j(x_t)\; Q_j(z)$$

valid for z>1, permit us to write

$$T^+(t,j) = \left[\bar{a}(t) - \frac{1}{\pi}\int_{x_o}^{\infty} \frac{A^+(t,z)}{z}\, dz\right] \delta_{jo}$$

$$+\frac{1}{\pi}\int_{x_o}^{\infty} A^+(t,z)\; Q_j(z)\; dz \qquad \text{(D5)}$$

and $$T^-(t,j) = \frac{1}{3}\left[\bar{b}(t) - \frac{1}{\pi}\int_{x_o}^{\infty} \frac{A^-(t,z)}{z^2}\, dz\right] \delta_{j1}$$

$$+\frac{1}{\pi}\int_{x_o}^{\infty} A^-(t,z)\; Q_j(z)\; dz. \qquad \text{(D6)}$$

These expressions are written in this form to make explicit the remark that for integer $j>0$ in T^+, and integer $j>1$ in T^-, the partial wave amplitudes are given simply by the integrals over A^+Q_j and A^-Q_j, while for $j=0$ and $j=1$ they are given by the subtraction term $\bar{a}$ and $\bar{b}$. Indeed, if A^+ is proportional to z (as would be the case for a constant total cross section) the integral over A^+Q_j does not exist for $j\leq 1$ and in particular at $j=0$, since $Q_j(z)\propto z^{-j-1}$ as $z\to\infty$.

Let us look at (D5) first. The term $\int A^+Q_j dz$ can be analytically continued in j to all values of j to the right of j_o, that is all values of j such that $\text{Re}\, j>j_o$, where $A^+\sim z^{j_o}$ for large z. Thus in the right-hand part of the complex j plane, $T^+(t,j)$ is an analytic function of j. We may now attempt to continue this function further to the left in the j-plane. (The explicit representation (D5), of course, fails since the integral ceases to exist.) Two possibilities exist. If the δ_{jo} term is in fact present, we can see that the physical amplitude defined by (D5) cannot coincide with our continuation from the right-hand j-plane. Conversely, if the coefficient of δ_{jo} vanishes, it is possible that the physical amplitude does coincide with the continuation. We shall assume that this second alternative in fact obtains, and that we have therefore defined an analytic continuation of $T^+(t,j)$, away from the set of even integers at which the physical partial wave amplitude is defined, into the complex j plane. We shall assume a similar statement obtains for $T^-(t,j)$ as well.[8]

Let us now return to the partial wave expansion (D5), and rewrite it using a Sommerfeld-Watson transformation.[9]

$$
\begin{aligned}
T^+(s,t) &= \sum_{j\ \mathrm{even}} (2j+1)\ P_j(x_t)\ T^+(t,j) \\
&= \frac{\pi}{2\pi i} \int_{C_1} \frac{dj\,(2j+1)}{\sin \pi j}\ P_j(-x_t)\ T^+(t,j).
\end{aligned}
\tag{D7}
$$

Where the contour C_1 surrounds the even integers, not the odd ones, as shown in fig. D1. Thus the singularities at $j=1,3\ldots$ due to $\sin\pi j$, as well as possible singularities of $T^+(t,j)$ at any j, lie outside C_1.

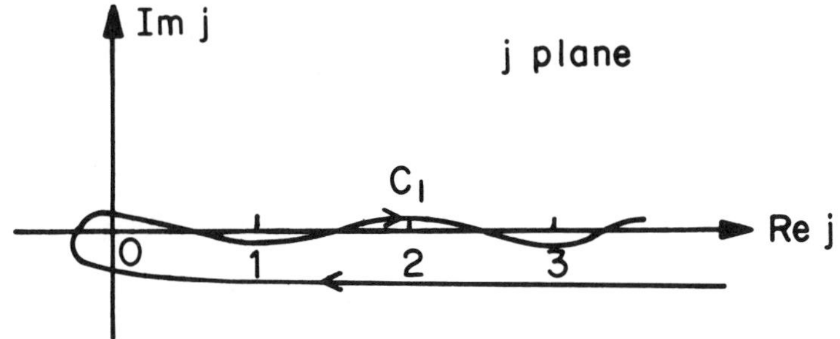

Fig. D1

Now let's sweep the contour C_1 outwards to a contour C_2, which is a vertical line just to the right of the right-hand-most singularity of $T^+(t,j)$, together with the remains of C_1 to the left of this line. C_2 is shown in fig. D2.

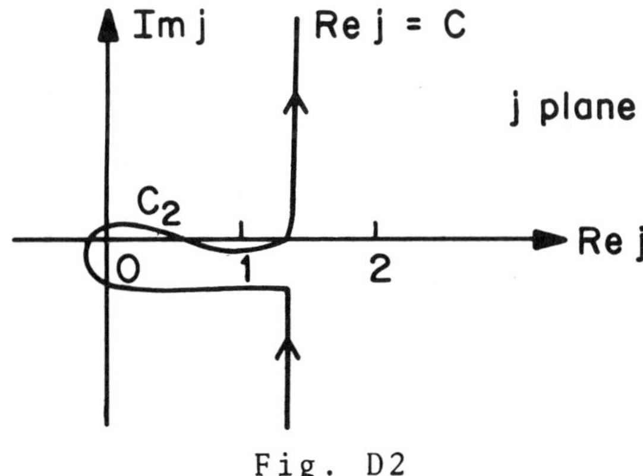

Fig. D2

In the process of this sweeping out, we pick up contributions from poles of the integrand at j=odd integers, due to the $\sin\pi j$ factor. We have

$$T^+(s,t) = \frac{\pi}{2\pi i} \int_{C_2} \frac{dj\,(2j+1)}{\sin \pi j} P_j(-x_t)\, T^+(t,j)$$

$$- \sum_{\substack{j \text{ odd} \\ j > j_{max}}} (2j+1)\, P_j(x_t)\, T^+(t,j). \quad \text{(D8)}$$

The terms at odd integer j in (D8) have no physical significance; since T(s,t) is proportional to $T^+(s,t)+T^+(u,t)$, these all cancel out. We are therefore free to ignore them.

The point of the Sommerfeld-Watson transformation was, as usual, to rewrite the partial wave expansion in a way which permits us to study the limit $x_t \to \infty$; i.e., $s \to \infty$. We could not do this with (D4) because the partial wave expansion does not converge outside the Lehmann ellipse[9]; we can do it with (D8).

When $s \to \infty$, $x_t \to s/2q_t^{\,2}$, and (D8) becomes (we ignore the $\sum_{j \text{ odd}}$ terms)

$$T^+(s,t) \to \frac{\pi}{2\pi i} \int_{C_2} \frac{(2j+1)\,dj}{\sin \pi j} \left(-\frac{s}{2q_t^{\,2}}\right)^j \frac{2^j}{\sqrt{\pi}} \cdot$$

$$\cdot \frac{\Gamma(j+\frac{1}{2})}{\Gamma(j+1)}\, T^+(t,j). \quad \text{(D9)}$$

For convenience let's absorb all the irrelevent factors into the partial wave amplitude by defining

$$\tilde{T}^+(s,t) = \pi \frac{2^j}{\sqrt{\pi}} \frac{\Gamma(j+\frac{1}{2})}{\Gamma(j+1)} \left(\frac{s_o}{2q_t^{\,2}} \right)^j T^+(t,j) \quad \text{(D10)}$$

where s_o is arbitrary. We can then write

$$T^+(s,t) \to \frac{1}{2\pi i} \int_{C_2} \frac{dj}{\sin \pi j} \left(-\frac{s}{s_o} \right)^j \tilde{T}^+(t,j). \quad \text{(D11)}$$

We note that for $t<4m_A^{\,2}$, $\tilde{T}^+(t,j)$ is real. Nevertheless, T^+ is still complex because of the phase in $(-s)^j$. Its imaginary part is

$$A^+(s,t) \to \frac{1}{2\pi i} \int_{C_2} dj \left(\frac{s}{s_o} \right)^j \tilde{T}^+(t,j). \quad \text{(D12)}$$

In this equation we can now forget the tail on the C_2 contour, since the $\sin\pi j$ is no longer present. Thus C_2 is just a vertical line at $\mathrm{Re}\,j=C$.

Eq. (D12) is in the form of a Mellin transform, though we must remember that it is valid only for s, that is for x_t, large. We may, however, define a function $\tilde{A}^+(s,t)$ by

$$\tilde{A}^+(s,t) = \frac{1}{2\pi i} \int_{C-i\infty}^{C+i\infty} dj \left(\frac{s}{s_o} \right)^j \tilde{T}^+(t,j) \quad \text{(D13)}$$

for all x_t. This function $\tilde{A}$ coincides with the physical $A^+(s,t)$ when s is large. Since $\tilde{A}$ is now defined for all s, the Mellin transform can be inverted to obtain

$$\tilde{T}^{+}(t,j) = \int_{s_o}^{\infty} d\left(\frac{s}{s_o}\right)\left(\frac{s}{s_o}\right)^{-j-1} \tilde{A}^{+}(s,t). \tag{D14}$$

These equations, (D13) and (D14), permit us at long last to translate freely between the s,t and the t,j languages.

Let's summarize the final statement: (We henceforth drop the ~ sign on the partial wave amplitudes.) We may write the entire amplitude at large s as

$$T(s,t) = \frac{1}{2\pi i}\int_{-i\infty}^{i\infty} dj \left\{ -\frac{(-s)^{j}+s^{j}}{\sin \pi j} T^{+}(t,j) - \frac{(-s)^{j}-s^{j}}{\sin \pi j} T^{-}(t,j) \right\}$$

which may be rewritten in the form

$$T(s,t) = \frac{1}{2\pi i}\int_{-i\infty}^{i\infty} dj \ s^{j} e^{-i\pi j/2} \left\{ -\frac{T^{+}(t,j)}{\sin \pi\frac{j}{2}} + i \frac{T^{-}(t,j)}{\cos \pi\frac{j}{2}} \right\} .$$

The absorptive part is

$$A(s,t) = \frac{1}{2\pi i}\int_{-i\infty}^{i\infty} dj \ s^{j} \left\{ T^{+}(t,j)+T^{-}(t,j) \right\}.$$

If we define A(s,t) at all s through this equation, it may be inverted to give

$$T^{+}(t,j)+T^{-}(t,j) = \int_{1}^{\infty} ds\ s^{-j-1}\ A(s,t).$$

Next, let's do some examples:

(i) Suppose $T^{+}(t,j)$ has a single pole at $j=\alpha(t)$; say $T^{+}(t,j) = \frac{\beta(t)}{j-\alpha(t)}$ plus other stuff. Then $\tilde{T}^{+}(t,j)=\frac{\tilde{\beta}(t)}{j-\alpha(t)}$ plus other stuff. Therefore, from (D11) we see that

$$T^{+}(s,t) = -\frac{\tilde{\beta}(t)}{\sin \pi\alpha(t)} \left(-\frac{s}{s_o}\right)^{\alpha(t)} + \ldots$$

and hence

$$T(s,t) = -\frac{\tilde{\beta}(t)}{\sin \pi\alpha(t)}\ \frac{1}{2} \left[\left(-\frac{s}{s_o}\right)^{\alpha(t)} + \left(\frac{s}{s_o}\right)^{\alpha(t)}\right] + \ \ldots$$

Similarly, from (D12) we have

$$A^{+}(s,t) = \tilde{\beta}(t) \left(\frac{s}{s_o}\right)^{\alpha(t)}$$

so that

$$A(s,t) = \frac{1}{2}\ \tilde{\beta}(t) \left(\frac{s}{s_o}\right)^{\alpha(t)}$$

Therefore: Single poles in j correspond to powers in s.

(ii) There is one exception to the previous remark; suppose $T^+(t,j)$ has a single pole in j at an odd integer, say at j=1. This power happens to coincide with the point at which $\sin\pi j$ vanishes, hence the integrand in (D11) actually has a double pole at j=1. Let

$$T^+(t,j) = \frac{\beta(t)}{j-1} + \ldots$$

then $$\tilde{T}^+(t,j) = \frac{\tilde{\beta}(t)}{j-1} + \ldots$$

Hence (D11) tells us that

$$T^+(s,t) = \frac{\tilde{\beta}(t)}{\pi} \frac{s}{s_o} \ln\left(-\frac{s}{s_o}\right)$$

while (D13) states that

$$A^+(s,t) = \tilde{\beta}(t) \left(\frac{s}{s_o}\right).$$

Thus a simple pole at this wrong signature integer corresponds to an $s \ln s$ behavior in $T^+(s,t)$ but a single power s in $A^+(s,t)$.

The total amplitude T(s,t), of course, looks like

$$T(s,t) = \frac{\tilde{\beta}(t)}{2\pi}\left(\frac{s}{s_o}\right)\left(\ln\left(-\frac{s}{s_o}\right) + \ln\left(\frac{s}{s_o}\right)\right) + \ldots$$

$$= \frac{i\tilde{\beta}(t)}{2}\left(\frac{s}{s_o}\right) + \ldots$$

$$= iA(s,t).$$

(iii) As another example, suppose $T^+(t,j)$ has a branch point at $j=\alpha_c(t)$, which we may represent by writing

$$T^+(t,j) = \frac{1}{\pi}\int_{-\infty}^{\alpha_c(t)} \frac{\beta(t,j')dj'}{j'-j} + \ldots$$

Then we have a similar expression for T^+ and we find from (D11) that

$$T^+(s,t) = \frac{1}{\pi}\int_{-\infty}^{\alpha_c(t)} dj' \frac{\tilde{\beta}(t,j')}{\sin\pi j'}\left(-\frac{s}{s_o}\right)^{j'} + \ldots$$

Similarly we find from (D13) that

$$A^+(s,t) = \frac{1}{\pi}\int_{-\infty}^{\alpha_c(t)} dj' \, \beta(t,j')\left(\frac{s}{s_o}\right)^{j'}.$$

For example, if the cut is of the form $(j-\alpha_c(t))^\nu$, then β and $\tilde{\beta}$ behave like $(j-\alpha_c(t))^{\nu+1}$ and we find $A^+(s,t)$ behaving like $(s/s_o)^{\alpha_c(t)} \cdot (\ln s/s_o)^{-\nu-1}$. If the cut is logarithmic, of the form $\ln(j-\alpha_c(t))$, then A^+ behaves like $(s/s_o)^{\alpha_c(t)} \cdot (\ln s/s_o)^{-1}$.

All the same results apply to $T^-(t,j)$ as well.

In this way, the following table can be constructed, connecting various kinds of j-plane behaviors of $T^\pm(t,j)$ and s behaviors of A(s,t):

j Plane	s Plane
$\frac{1}{j-\alpha}$	s^α
$(j-\alpha)^\nu$	$s^\alpha(\ell n\ s)^{-\nu-1}$
$\ell n\ (j-\alpha)$	$s^\alpha/\ell n\ s$
$[\ell n\ (j-\alpha)]^\nu$	$s^\alpha(\ell n\ \ell n\ s)^{\nu-1}/\ell n\ s$

The connection between the s,t language and the t,j language permits us to exploit another basic principle to constrain the possible asymptotic behaviors. This principle is t-channel unitarity. We know that for t above elastic but below inelastic threshold, we have

$$T^\pm(t,j)-T^\pm(t,j^*)^* = 2i\rho(t)T(t,j)T^*(t,j^*) \tag{D15}$$

where $\rho(t) = \frac{1}{16\pi}\sqrt{\frac{t-4m^2}{t}}$.

This relation is evidently true at even integer j for T^+ and odd integer j for T^-, since there $T^\pm$ coincides with the physical partial wave amplitudes. It is also true in the form stated for the continued amplitudes.[10)]

Eq. (D15) immediately tells us the following: $T^{\pm}(t,j)$ cannot contain, in this region of t, any singularity at a real value of j at which the amplitude is infinite - that is, any hard singularity.[11) For example, suppose $T^{+}(t,j)$ had a pole at $j=\alpha$ with α real. Then the left-hand side of (D15) has a single pole while the right-hand side has a double pole, which is impossible. The same argument applies to a hard branch point. A soft branch point at a real value of j is, in contrast, allowed. Also allowed is any moving singularity at a complex value of j.[12)

Let's first work on poles. The normal situation of a moving pole causes no difficulty. A pole at $j=\alpha(t)$ may be real for $t<4m_A^2$, but then $\alpha(t)$ becomes complex for $t>4m_A^2$. In fact, from (D15) we can easily see that just above threshold, we must have $\text{Im}\alpha(t)=\rho(t)/\beta(t)$, where β is the residue. A fixed pole is different. If for $t<4m_A^t$ there is a pole at j=1, say, independent of t, then analyticity suggest that this pole is at j=1 for $t>4m_A^2$ as well. But this is impossible, so we are led to forbid fixed poles. Therefore, we are led to forbid asymptotic behaviors like $A(s,t)\to\beta(t)s$, for example.

More generally, an asymptotic behavior like $A(s,t)\to\beta(t)F(s)$ corresponds to a fixed j plane singularity: Evidently $T(t,j)=\beta(t)F(j)$. The singularity must then (if real) be a soft branch point - e.g., $T(t,j)\sim(j-\alpha)^{\nu}$ where α is real and ν is positive. Returning to the s-plane, this says

$A(s,t)\to\beta(t)s^{\alpha}/(\ell n)^{\nu+1}$, so that total cross sections must vanish faster than $1/\ell n\ s$.

An escape from these restrictions is possible, but requires a very clever conspiracy of j-plane effects. Suppose we take the j=1 fixed pole example. We want this pole to exist for $t<4m_A^2$, but to disappear for $t>4m_A^2$. The convenient thing to do is to shunt it off onto an unphysical sheet of the j-plane. For this to be possible there must be a convenient j-plane branch point passing through j=1 at $t=4m_A^2$. Furthermore, since phase space is proportional to $\sqrt{t-4m_A^2}$, it must be a square root branch point.[13)]

An explicit example of such a situation will serve to make these points clear. Suppose we have the form

$$T^{+}(t,j)=$$

$$=\frac{A(t,j)/(j-1)}{1-\dfrac{A(t,j)}{16\pi^2}\displaystyle\int_{-\infty}^{\alpha_c(t)}\sqrt{\frac{\alpha_c(t)-j'}{\bar{\alpha}_c(t)-j'}}\ \frac{B(t,j')dj'}{(j'-1+i\varepsilon)(j'-j)}}. \tag{D16}$$

The contour of integration passes above the pole at j'=1. The physical sheet is defined by Imj>0, and the physical amplitudes by letting Imj→0+. We assume, further, that $\alpha_c(4m_A^2)=1$, say $\alpha_c(t)=1+\alpha'(t-4m_A^2)$. Then $\bar{\alpha}_c(t)$ is defined to be $1+\alpha' t$. Finally, we must require that A and B are arbitrary functions real in the region between elastic and inelastic thresholds, with B(t,1)=1.

All of this serves to arrange our magic situation as follows:
(i) for $t<4m_A^2$, $\alpha_c(t)<1$ the integral in (D16) has no singularities, and is real. Thus $T^+ \alpha\ 1/(j-1)$.
(ii) as soon as $t>4m_A^2$, $\alpha_c(t)>1$ and now the singularities of the integral are relevant. We have, in fact,

$$T^{-1}(t,j) - T^{-1}(t,j^*)^* =$$

$$= -\frac{j-1}{16\pi^2}\int_{-\infty}^{\alpha_c}\sqrt{\frac{\alpha_c-j'}{\bar{\alpha}_c-j'}}\ \frac{B(t,j')dj'}{j'-j}\left(\frac{1}{j'-1+i\varepsilon}-\frac{1}{j'-1-i\varepsilon}\right)$$

$$= \frac{j-1}{16\pi}\sqrt{\frac{\alpha_c-1}{\bar{\alpha}_c-1}}\ \frac{B(t,1)}{1-j} = -\frac{1}{16\pi}\sqrt{\frac{t-4m^2}{t}}$$

$$= -\rho(t) \tag{D17}$$

since $B(t,1)=1$.

Thus the pole has vanished and unitarity is satisfied. It has disappeared onto the unphysical sheet of the square root cut.

The dynamical origin of such a square root cut remains a mystery. Among known j-plane singularities, the most similar object is the Reggeon-particle cut, which is probably a square root cut and passes through any two particle threshold at $j=J_1+J_2-1$, where J_1 and J_2 are the spins of the two particles making up the threshold.[14] If this value were j=1 instead, this cut would be a likely candidate.

A final point of interest has to do with the extension of the above discussion to many channels. For a moving Regge pole, the unitarity relation (D15) can be used to show that the residue function factors;[15)] this is to be expected physically in any case since if particles lie on the trajectory, their coupling constants must factor. More generally, however, any hard singularity which occurs in a single eigenchannel of the S-matrix has a factoring coefficient. Thus factorization is essentially a matter of choice; any singularity can be chosen to factor, but only for the case of moving poles having physical particles does physics demand factorization.

E

Two-Body Intermediate States

We have frequently in the text found occasion to refer to the two-body intermediate state contribution to an amplitude. Examples are: The two-particle s-channel discontinuity of a Feynman diagram, or the two-body intermediate state contribution in the s-channel unitarity relation. The following calculations are written for the latter case. We wish here to present briefly the relevant arithmetic leading to some of the formulae we make use of.

Let us write

$$A_2(s,t) = \frac{1}{2}\int\frac{d^3q_1}{(2\pi)^3}\int\frac{d^3q_2}{(2\pi)^3}\,\frac{1}{2\omega_1}\,\frac{1}{2\omega_2}(2\pi)^4\cdot$$

$$\cdot\ \delta^{(4)}(P-q_1-q_2)T(s,t_1)T^*(s,t_2) \quad \text{(E1)}$$

where $P^2=s, (p_1-p_1')^2=t$, $(p_1-q_1)^2=t_1$, and $(p_1'-q_1)^2=t_2$. We shall concern ourselves only with

$s \gg t, t_1, t_2, \mu^2$ where we take $\omega_1 = \sqrt{\vec{q}_1^{\,2} + \mu^2}$ and $\omega_2 = \sqrt{\vec{q}_2^{\,2} + \mu^2}$. We take $p_1^2 = p_2^2 = p_1'^2 = p_2'^2 = \mu^2$ as well.

The delta function simplifies the integral to

$$A_2(s,t) = \frac{1}{64\pi^2} \int d\Omega_{\hat{q}_1} \; T(s,t_1) T^*(s,t_2). \tag{E2}$$

If we define $x_1 = \hat{p}_1 \cdot \hat{q}_1$, $x_2 = \hat{p}_1' \cdot \hat{q}_1$ and $x = \hat{p}_1 \cdot p_1'$, the integration over angles θ and ϕ of $\hat{q}_1$ can be changed to

$$A_2(s,t) = \frac{1}{32\pi^2} \int \frac{dx_1 \; dx_2}{\sqrt{1 - x^2 - x_1^2 - x_2^2 + 2xx_1x_2}} \cdot T(s,t_1) T^*(s,t_2). \tag{E3}$$

The range of integration is over those x_1 and x_2 for which the argument of the square root is positive. Since $t = -(s/2)(1-x)$ and $t_{1,2} = -(s/2)(1-x_{1,2})$, we can next change variables to obtain

$$A_2(s,t) = \frac{1}{16\pi^2 s} \iint \frac{dt_1 \; dt_2}{\sqrt{-\lambda}} \; T(s,t_1) T^*(s,t_2) \tag{E4}$$

where
$$\lambda(tt_1t_2s) = t^2 + t_1^2 + t_2^2 - 2tt_1 - 2tt_2 - 2t_1t_2 + \frac{4tt_1t_2}{s - 4\mu^2}$$
$$\xrightarrow[s\to\infty]{} t^2 + t_1^2 + t_2^2 - 2tt_1 - 2tt_2 - 2t_1t_2$$
$$\equiv \lambda(tt_1t_2) \tag{E5}$$

as $s\to\infty$. The integration, again, is over t_1 and t_2 such that $\lambda<0$.

There are two magic formulas for this kinematic square root, namely:

$$\frac{1}{\sqrt{-\lambda}} = \frac{\pi}{s}\sum_j (2j+1)P_j(x_1)P_j(x_2)P_j(x) \tag{E6}$$

and[16)]

$$\frac{1}{\sqrt{-\lambda}} = \frac{\pi}{2}\int_0^\infty bdb\ J_o(b\sqrt{-t_1})J_o(b\sqrt{-t_2})J_o(b\sqrt{-t}). \tag{E7}$$

It is sometimes of interest to rewrite eq. (E4) in terms of the t-channel partial wave amplitude corresponding to the various amplitudes. If we define, as usual,

$$T(s,t) = \frac{1}{2\pi i}\int_{-i\infty}^{i\infty} dj\ s^j\ e^{-i\pi j/2}\left\{-\frac{T^+(t,j)}{\sin\frac{\pi j}{2}} + i\,\frac{T^-(t,j)}{\cos\frac{\pi j}{2}}\right\} \tag{E8}$$

and

$$A_2(s,t) = \frac{1}{2\pi i}\int_{-i\infty}^{i\infty} dj\ s^j\left\{T_2^+(t,j)+T_2^-(t,j)\right\} \tag{E9}$$

then eq. (E4) tells us that

$$T_2^+(t,j) + T_2^-(t,j) =$$

$$= \frac{1}{16\pi^2}\iint \frac{dt_1 dt_2}{\sqrt{-\lambda}} \frac{1}{2\pi i}\int_{-i\infty}^{i\infty} dj_1 \frac{1}{2\pi i}\int_{-i\infty}^{i\infty} dj_2 \cdot$$

$$\cdot \frac{1}{j-j_1-j_2+1} e^{-\frac{i\pi}{2}(j_1-j_2)} \left(\frac{T^+(t_1,j_1)}{\sin \frac{\pi j_1}{2}} \right.$$

$$\left. - i \frac{T^-(t_1,j_1)}{\cos \frac{\pi j_1}{2}} \right) \left(\frac{T^+(t_2,j_2)^*}{\sin \frac{\pi j_2}{2}} + i \frac{T^-(t_2,j_2)^*}{\cos \frac{\pi j_2}{2}} \right) . \tag{E10}$$

REFERENCES

1. A.H. Mueller, Phys. Rev. D2, 2963 (1970).
2. C.I. Tan, Phys. Rev. D4, 2412 (1972).
   K.E. Cahill and H.P. Stapp, Phys. Rev. D6, 1007 (1972).
   J.C. Polkinghorne, Nuovo Cim. 7A, 555 (1972).
3. M. Froissart, Phys. Rev. 123, 1053 (1961).
4. A. Martin "Strong Interactions," Scottish Universities' Summer School, 1963, Oliver and Boyd, London, p. 105.
5. I. Ya. Pomeranchuk, Soviet Physics JETP 3, 307 (1956).
6. See any review of Regge theory such as those given in reference 1 of part II.
7. See for example "Higher Transcendental Functions, Vol. I," Erdelyi et al., (McGraw-Hill, New York 1954) for properties of Legendre functions.
8. G.F. Chew, "S-Matrix Theory of Strong Interactions," (W.A. Benjamin, Inc., New York, 1961).
9. T. Regge, Nuovo Cim. 14, 951, 1960 (1959).
10. H. Lehmann, Nuovo Cim. 10, 579 (1958).

11. V.N. Gribov, Soviet Physics JETP 14, 1395 (1962).
A.O. Barut and D. Zwanziger, Phys. Rev. 129, 974 (1962).
K. Bardacki, Phys. Rev. 127, 1832 (1962).
E.J. Squires, Nuovo Cim. 25, 242 (1962).
R. Oehme, Phys. Rev. Letters 9, 358 (1962).
12. V.N. Gribov, Nucl. Phys. 22, 249 (1961).
R. Oehme, Phys. Rev. Letters 18, 1222 (1967).
13. R. Oehme, "Strong Interactions and High Energy Physics," Oliver-Boyd, Edinburgh, 1962, p. 129.
14. J. Schwarz, Phys. Rev. 162, 1671 (1967).
15. M. Gell-Mann, Phys. Rev. Letters 8, 263 (1962).
V.N. Gribov and I. Ya. Pomeranchuk, Phys. Rev. Letters 8, 343 (1962).
16. Erdelyi et al., "Higher Transcendental Functions," Vol. II, McGraw-Hill, New York, 1954.
17. R.J. Eden, High Energy Collision of Elementary Particles, Cambridge University Press, 1967.

Index